Springer Tracts in Natural Philosophy

Volume 33

Edited by C. Truesdell

Springer Tracts in Natural Philosophy

Harley Cohen   Robert G. Muncaster

The Theory of Pseudo-rigid Bodies

With 12 Illustrations

Springer-Verlag
New York Berlin Heidelberg
London Paris Tokyo

Harley Cohen
Department of Civil Engineering
University of Manitoba
Winnipeg, Manitoba, Canada R3T 2N2

Robert G. Muncaster
Department of Mathematics
University of Illinois at Urbana—Champaign
Urbana, Illinois 61801, USA

Mathematics Subject Classification (1980) 70E15

Library of Congress Cataloging-in-Publication Data
Cohen, Harley.
  The theory of pseudo-rigid bodies.
  (Springer tracts in natural philosophy; v. 33)
  Bibliography: p.
  Includes index.
  1. Dynamics, Rigid. 2. Deformations (Mechanics)
3. Continuum mechanics. I. Muncaster, R. G.
II. Title. III. Series.
QA861.C64 1988   531'.38   87-28477

9 8 7 6 5 4 3 2 1

ISBN-13: 978-1-4613-9591-1   e-ISBN-13: 978-1-4613-9589-8
DOI: 10.1007/978-1-4613-9589-8

For
Estelle and Nancy

Preface

This monograph concerns the development, analysis, and application of the theory of pseudo-rigid bodies. It collects together our work on that subject over the last five years. While some results have appeared elsewhere, much of the work is new. Our objective in writing this monograph has been to present a new theory of the deformation of bodies, one that has not only a firm theoretical basis, but also the simplicity to serve as an effective tool in practical problems. Consequently, the main body of the treatise is a multifaceted development of the theory, from foundations to explicit solutions to linearizations to methods of approximation. The fact that this variety of aspects, each examined in considerable detail, can be collected together in a single, unified treatment gives this theory an elegance that we feel sets it apart from many others.

While our goal has always been to give a complete treatment of the theory as it now stands, the work here is not meant to be definitive. Theories are not entities that appear suddenly one day and thereafter stand as given. Rather, they must mature and grow with time and experience. Our development is more correctly a beginning, tempting others to explore, appraise, and modify its features so as to produce something better. We hope that the foundations laid here are sufficiently firm to endure; but we also hope that, in both its simplicity and its rich variety of applications, the theory will encourage others to use it in their own work.

Without the continued interest and encouragement of CLIFFORD TRUESDELL, as a teacher and as a colleague, this work would not have been attempted. He has allowed us the freedom to collect together our results in this compact, concise, and complete presentation, and for this we owe a sincere debt. Several others, both friends and colleagues, have contributed through important discussions and criticisms of early drafts of the monograph, and to these we also offer our thanks. They are TIM HEALEY, CHI-SING MAN, GEAROID MACSITHIGH, and JOHN REED.

We also acknowledge the generous financial support of the U.S.

National Science Foundation and the Natural Sciences and Engineering Research Council of Canada.

Finally, our sincerest debt goes to NANCY MUNCASTER for the composition of the manuscript. She typed the first, second, third, ..., and final drafts, and we gratefully acknowledge her skills and infinite patience.

Winnipeg, Manitoba   Harley Cohen
Urbana, Illinois   Robert G. Muncaster

Contents

CHAPTER 1

Prolegomena to the Theory

The theory of pseudo-rigid bodies focuses on the large-scale motions of deformable bodies. It provides a convenient framework, like classical rigid-body mechanics, for the analysis of gross changes in the position and orientation of a body. As such it represents a generalization of that classical theory. At the same time, the theory of pseudo-rigid bodies concerns deformation; treating changes in the shape of a body by use of certain gross measures of strain. As such it represents a restriction, or coarse version, of many theories now commonplace in continuum mechanics. Between these two extremes, the modern and the classical, the theory of pseudo-rigid bodies takes a middle road, focusing on problems that exhibit a high degree of interplay between deformation and rigid-body motion.

This chapter gives an overview of the different perspectives of our subject. We begin with a discussion of the place the theory occupies in the historical development of continuum mechanics. In retrospect, this theory pulls together and solidifies a number of ideas put forward by the past masters, ideas that have arisen in a natural progression of thought and which, at least for pseudo-rigid bodies, we have been able to give a precise development. We look at these historical perspectives in Section 1(i). Next, we examine the role that the theory plays as a study in rational mechanics. This study, we feel, is a specimen for the student of the different ingredients that can, and perhaps should, be included in any study of a new theory. The different steps in this study form the main body of the monograph, and they are summarized in Section 1(ii). In the course of our writing, a third theme has evolved, one that has been treated only diffusely in the literature despite its importance for future work in continuum mechanics. This is the study of *cross-theory comparison*. In many ways, the theory of pseudo-rigid bodies is one of the simplest in a hierarchy of theories, some of great age, others now being developed, and still others only yet in the imagination. We present here an analytical technique for the comparison of different theories and then use it to show how the theory of pseudo-rigid bodies can be appraised through comparisons both with simpler theories to which it should reduce by specialization and with more complex theories for which it should serve as a specialization. This method of comparative analysis encompasses new questions and new techniques that are developed here in sufficient detail that they speak to this hierarchy of theories. We highlight the different parts of the study of cross-theory comparison in Section 1(iii). Thereafter, we move to the formal development of the theory of pseudo-rigid bodies beginning in Chapter 2.

(i) Historical Perspectives

Continuum mechanics draws a clear distinction between balance laws as statements of general principles and constitutive relations as definitions of idealized material response. This distinction can be traced back to EULER,[1] who had clear definitions of the rigid body and the perfect fluid. NEWTON was certainly aware of these two materials; he also had an appreciation of the viscous fluid. The definition of an elastic material had to await the development of new concepts, and it remained for CAUCHY to first formulate its linear constitutive relation.

The theory of elasticity and the mechanics of rigid bodies each followed its own path of development, both reaching sophisticated levels by the twentieth century. Initially, the vast majority of work on elastic materials was restricted to the linear theory. Not until the middle of the present century did nonlinear elasticity begin to yield to analysis and attract attention. Similar successes were seen for nonlinear viscous fluids. These successes brought about a resurgence of interest in the overall subject of continuum physics, and rational mechanicists and engineering scientists took up the study with new vigor. Specialized theories were united within one framework, and the separate status of general laws and constitutive relations was reclarified. At the same time, rigid-body mechanics, although a theory of a special type of continuum, appeared instead to be abandoned to physics and applied mathematics for further study.

Early investigations of elastic bodies were mainly concerned with the formulation of special theories for thin and slender bodies. Rods, strings, shells, and membranes—the basic elements of structural engineering— provided the main impetus for research. Initially, these *structural* theories lacked the unifying framework that the general three-dimensional equations of classical linear elasticity would eventually provide; thereafter

---

[1] This section provides a skeletal historical framework from which to view our subsequent developments. It is not a precise and detailed history. For an authoritative discussion of the history of constitutive relations, we refer the reader to TRUESDELL [1980].

developed the tradition of *deriving* structural theories, by various scalings and approximations, from the three-dimensional theory.[2]

E. COSSERAT and F. COSSERAT [1909], in a monumental work, provided the foundations for a new approach to the mechanics of thin and slender bodies. Like the earlier research, their approach permitted the formulation of structural theories without recourse to an independent three-dimensional theory. In contrast, however, the COSSERAT brothers worked within a *structured* framework rather than by more or less *ad hoc* theorizing. In the sequel, we shall refer to mechanical theories for special bodies as *derived* or *direct*, depending on whether or not they are obtained from a parent three-dimensional theory.

The COSSERATS, following an idea of DUHEM [1893], introduced the *directed continuum*. This is a classical continuum[3] with a set of rigid vectors called *directors* attached to each of its points. The classical continuum had traditionally served as a model for a *body* in mechanics. The directed continuum generalized this idea, providing a model for what we call here an *extended body*. The extra kinematic freedom associated with the directors could be used to model aspects of material structure inaccessible in terms of the classical body. Strings and membranes had satisfactorily been modeled, respectively, by one- and two-dimensional continua—curves and surfaces in space. With the extra structure provided by the directors, rods and shells were modeled as directed curves and surfaces: the directors captured variations in the cross-section and thickness, respectively.

The work of the COSSERATS lay incomplete, unheralded, and with few exceptions unnoticed. In 1958 it was recalled, modernized, and extended to continua with *deformable* directors in a major study by ERICKSEN and TRUESDELL. Possibly more important, their work introduced the concept of the directed continuum to researchers in mechanics. The timing could not have been more propitious. The aforementioned successes in the analysis of nonlinear theories of simple materials prompted those cultivating structural theories to seek nonlinear results for rods and shells. Their initial approach, however, was hampered rather than simplified by past experience. Tradition guided them to seek theories that were derived from a nonlinear three-dimensional theory. Unfortunately, geometric and kinematic complexities surrounded this approach and led instead to a maze of insurmountable difficulties. Success ultimately lay in a new direction—the method of the COSSERATS based on the concept of the directed continuum.

The framework set down by ERICKSEN and TRUESDELL [1958] has

---

[2] For further details we refer the reader to the historical introduction given by LOVE [1927] in his treatise on elasticity.

[3] The classical continuum is a three-dimensional differentiable manifold that can be covered by one coordinate chart.

been used as the basis of a number of direct theories. COHEN and
DESILVA [1966] adopted it for their analysis of hyperelastic directed
surfaces, while COHEN [1966] used it for hyperelastic directed curves.
Nonlinear theories of rods and shells quickly followed. In essence, the
program set forth by the COSSERATS has been realized. The status of this
work to recent times, including extensive bibliographies, is described by
ANTMAN [1972] for rods and NAGHDI [1972] for shells.

At the same time, the treatises of ANTMAN and NAGHDI attest to
the fact that derived theories had not been abandoned. Indeed, now
that direct theories provided a path through the maze, the question of
deriving structural theories could be attacked with renewed vigor. Of
primary interest was the relation between the two theories—the direct
and the derived. TRUESDELL [1959] raised this question in terms of the
compatibility of solutions to a given boundary-value problem. He called
this the *consistency* problem. MUNCASTER [1984a], using the concept of
fine-coarse theory pairs, defined the notion of consistency in terms of the
relation between solutions of the derived (coarse) theory and those of the
parent (fine) theory. His notion of consistency, although less general than
TRUESDELL'S, was phrased in a precise mathematical form and led to
equations through which consistency could be addressed directly.

The COSSERATS' idea of a three-dimensional generalized continuum,
once introduced to continuum mechanicists by ERICKSEN and TRUESDELL,
gave rise to a wealth of study. Further aspects of three-dimensional generalized continua, especially those related with stress, are dealt with in
the article by TRUESDELL and TOUPIN [1960]. Other landmark papers are
those of TOUPIN [1964] and MINDLIN [1964], which developed theories
of elastic directed materials. A review of this early work may be found in
TRUESDELL and NOLL [1965], while some idea of its expanding scope can
be found in KRONER [1968]. A more recent guide to activity in this area
can be gleaned from the bibliography of the lecture notes edited by
BRULIN and HSIEH [1982].

Perhaps one of the surprises of this new work was the appearance of
concepts not inherent in the classical theory. The additional kinematic
freedom associated with the directors provided a need for generalized
stresses, new field equations, and corresponding new constitutive relations. Some aspects, such as nonsymmetric stress and couple stress, could
be anticipated from the work on theories of rods and shells; others, such
as microstress, higher-order stress, and microinertia, were new. A current
version of the equations of the three-dimensional directed continuum has
been provided by CAPRIZ and PODIO-GUIDUGLI [1976].

A recent innovation is the *zero-dimensional body with directors*.
MUNCASTER [1984b] formulated the basic equations of such a body as a
derived coarse theory, using as the fine theory the equations of nonlinear
elasticity. Independently, COHEN [1981] set down essentially the same
governing equations through a direct formalism. He called such bodies

pseudo-rigid. As we shall see later, these two theories—one derived and the other direct—are consistent in the sense of MUNCASTER'S original notion. The common feature that makes both theories especially attractive is the fact that each comprises a dynamical system with finitely many degrees of freedom.[4] Consequently, the governing equations are a system of ordinary differential equations and hence are generally more tractable than conventional initial-boundary-value problems for deformable media.

In this monograph, we refer to both the aforementioned theories as *the theory of pseudo-rigid bodies*. Our purpose is to formulate and apply it, in particular, in the case of an *elastic* pseudo-rigid body. The theory may therefore be viewed either as a generalization of the classical mechanics of a rigid body or as a restriction of the classical theory of an elastic body. It represents a modern melding of the two, and yet, as we shall see, it stands independent, with a life and character of its own.

(ii) Toward the Rational Mechanics of Pseudo-rigid Bodies

As a study in rational mechanics, this monograph is rather unusual. Some studies concentrate on laying foundations, with careful attention to kinematics and constitutive theory, and resulting in field equations for a new theory of material behavior. Other studies are centered within a well-accepted theory and focus on the analysis and interpretation of particular problems. Yet other studies aim at the development of schemes of linearization or successive approximation that permit the solution of classes of otherwise intractable problems. In the theory to be presented here, we explore all three of these perspectives. This is a testament, we feel, to our initial desire for a theory with both a firm foundation in continuum mechanics and practical value for solving specific problems.

As our historical development makes clear, the basic equations of motion for pseudo-rigid bodies may be arrived at in two different ways—by direct and derivation methods. Chapters 2 and 3 explore these respective approaches in detail. We begin with the direct approach to the theory, viewing this largely as a complete development of the foundations of the subject. Thereafter, we examine the derivation method, but we deviate from tradition both in its presentation and the interpretation of its results. Specifically, we develop this method not as an alternate route to the same equations of motion but rather as a study in cross-theory comparison. The basic question is not whether we can derive a theory of pseudo-rigid bodies from some parent in three dimensions, but rather,

---

[4] Related work by SLAWIANOWSKI [1974, 1975], although different in approach, arrives essentially at the same theory formulated by COHEN and MUNCASTER. Moreover, equivalent equations can be obtained from the paper by CAPRIZ and PODIO-GUIDUGLI [1976] by restricting attention to a single point (the microelement) of their field theory.

whether the child we have now produced by direct methods is a *good representation* of some three-dimensional parent.

The foundations of our theory are laid in Chapter 2. The objective is to obtain a theory that generalizes classical rigid-body mechanics by allowing for the effects of deformation. The discussion, generalizing that of COHEN [1981], is an axiomatic treatment in which a pseudo-rigid body is modeled as a directed continuum. We begin by introducing the concept of an extended event world and then define a pseudo-rigid body as a vector bundle that is viewed within this world through a class of placements. The kinematics of motion are then considered, with special attention to generalized measures of mass and momentum, and thereafter generalizations of CAUCHY's equations of balance are set down. In connection with these, we examine in detail the form and interpretation of external loads for a pseudo-rigid body. In order to make contact with more classical results, we next look back at the mechanics of rigid bodies in the light of our developments and we identify the angular momentum, and its equation of balance, for a pseudo-rigid body. This is followed by discussions of changes of frame and then constitutive relations. At this point, the basic equations of motion for a pseudo-rigid body are complete. Beginning in Chapter 4, we consider the solution of selected problems and the development of perturbation methods based on these equations.

The direct approach has the advantage of simplicity and clarity, presenting the essence of the subject without recourse to the complications of more general overriding theories. At the same time, however, it does not allow us to answer the question: how do our predictions for pseudo-rigid bodies bear on the motions of truly three-dimensional bodies? The answer to this question, which we regard as an integral part of our study, can only be approached through a comparative analysis—a comparison of the theory of pseudo-rigid bodies on the one hand and some three-dimensional continuum theory on the other. We present such a comparison, phrased in terms of the *consistency* of two theories, in Chapter 3. Our approach to the notion of consistency is an outgrowth of MUNCASTER's [1984a] original concept of fine–coarse theory pairs. The objective is to ask whether the theory of pseudo-rigid bodies can be viewed as a coarse version of three-dimensional nonlinear elasticity. We begin with an extended discussion of the ideas behind the notion of consistency. This leads us to develop the concept of a *subtheory*, both generally and as it applies to the theory of pseudo-rigid bodies. The concept of a subtheory entails the specification of a mapping called a *coarsifier*, through which direct comparisons between theories are possible. The requirement that a coarse theory be "consistent" with a fine theory delivers a precise analytical problem for the computation of a coarsifier for the pair. At this point, the framework for a theory of consistency is in place, and we consider then two issues to which such a framework might speak. The first is the matter of *inherited symmetry*. In particular, we ask the question: if fine and coarse theories are consistent, and if the fine theory satisfies the principle of material frame indifference, does the coarse theory inherit this same property? The second problem we consider is the development of a scheme of approximation for the coarsifier.

One way to appraise the quality of the theory of pseudo-rigid bodies is to compare it with other theories. However, in order to examine its practical value, we must turn to the problem of finding solutions to its basic equations. This is the subject of Chapter 4. To set the stage, we look first at *rigid motions* of elastic pseudo-rigid bodies, thereby providing a nonstandard view of the mechanics of rigid bodies. Next, employing semi-inverse techniques, we consider a series of different problems. The first class concerns *roto-deformations*—solutions that exhibit a strong interplay between rotation and deformation. Then we examine solutions in which rotation is absent—the *pure stretch* solutions. In order to illustrate a case in which both strain and rotation are present, but in which it is neither simple nor natural to separate the two effects, we treat next *simple shearing* of a pseudo-rigid body. Finally, we conclude the chapter with an analysis of *plane rolling* of an elastic pseudo-rigid lamina.

In Chapter 5, we examine Lagrangian and Hamiltonian formulations of the theory and so place our subject within the realm of an extensive body of well-known analytical results and techniques. A novel feature of the variational framework in this case stems from the fact that the configuration manifold for the theory is a *Lie group*. We begin by presenting the variational formulation in a traditional format but quickly rework and compare this treatment with a formulation on the general linear group in three dimensions. Next, we focus more directly on the rotational part of the deformation. The principle of material frame indifference leads us immediately to questions of *invariance groups*, *symmetry*, and *conservation laws*. We discuss these in a nonstandard way through a variational formulation on the orthogonal group in three dimensions.

Historically, the Lagrangian format has been especially convenient for the treatment of small oscillations and questions of stability. In the last two sections of Chapter 5, we extend this tradition to pseudo-rigid bodies by examining the linearized stability of *steady motions*. We do this in two different ways. The first is a specialized analysis in terms of components of angular velocity, thus permitting direct comparison with classical analyses of the stability of a spinning rigid body. The second is a treatment in terms of *Euler angles* and *ignorable coordinates*, through which general questions of stability can be addressed.

A rigid body is an idealization that is useful when the effects of deformation are small enough to be considered unimportant. No body, however, is truly rigid. Indeed, there are many problems, such as those related to the long-time motion of satellites, in which deformations are small but still important. Bodies of this type might be called *almost rigid*. In Chapter 6, we present a perturbation analysis of the motion of a pseudo-rigid body in the neighborhood of a rigid state. The goal, in particular, is to examine a problem of considerable practical interest: the *gyroscopic motion* of an almost-rigid body. The perturbation technique involves a two-time scale asymptotic representation of solutions to the basic equations. The leading approximation is a motion according to the mechanics of rigid bodies, and we take it to correspond to the force-free steady precession of an axially symmetric body. The second and third terms in the perturbation scheme are determined explicitly, and we use these to highlight the interplay between the effects of deformation and rotation and to predict changes in the gyroscopic rotation of a body owing to its flexibility.

(iii) Toward the Study of Cross-theory Comparison

The comparison of theories is a theme that recurs repeatedly throughout this monograph. It stems from our desire to be able to appraise the quality of the theory of pseudo-rigid bodies, both in the course of our development of the subject and also once the foundations had finally been set in place. Certainly an important test of any theory lies in the predictions it gives in its application to a variety of problems. This involves a direct comparison between theoretical results and what we view as *realistic* mechanical behavior. For the theory of pseudo-rigid bodies, the analyses we present in Chapters 4, 5, and 6 serve largely this type of appraisal. Here, however, we are interested in another way to appraise theory, namely, through comparisons with other established theories that speak to similar applications.

The theory of pseudo-rigid bodies is ideally suited to cross-theory comparisons. In one respect, this theory is meant to generalize the mechanics of rigid bodies. Thus, comparisons should be made with that theory. In another respect, pseudo-rigid bodies are meant to model three-dimensional deforming bodies, and this suggests that comparisons should be made with theories of three-dimensional continua. In this regard, we focus specifically at the theory of nonlinear elasticity. Finally, the theory of pseudo-rigid bodies is perhaps the simplest in the general hierarchy of theories of directed continua. Thus, here too comparisons should be examined.

In its simplest and most common form, the comparison of theories proceeds at a purely descriptive level, often phrased in terms of "similarities", "parallels", and "analogies." The first systematic and analytical studies that reflected a concern for cross-theory comparison can be found in the classical work on rods and shells. These emphasized the derivation of one theory from another, thereby imposing a strong comparison directly by construction (cf. ANTMAN [1972] and NAGHDI [1972]). Into this same category falls ENSKOG's celebrated work in gas dynamics in which continuum gas dynamics was derived from the kinetic theory of gases (cf. TRUESDELL and MUNCASTER [1980]). Motivated by the studies in gas dynamics and on rods and shells, MUNCASTER [1984a] formulated in abstract terms the analytical study of pairs of theories, one fine and the other coarse, with emphasis again on the derivation of one from the other. By extending, revising, and reinterpreting MUNCASTER's original ideas, we have formalized the comparison of theories through the notion of *consistency*. The details of this formulation are presented in Chapter 3 and have been summarized in Section 1(ii).

The cross-theory comparisons we develop fall into four different classes. The first consists of direct comparisons between the theory of pseudo-rigid bodies and the mechanics of rigid bodies. While analogies between the two arise continually, certain comparisons are developed in detail. In Section 2(v), we explore the sense in which rigid-body mechanics is a consistent subtheory of pseudo-rigid-body mechanics. Through this comparison, we arrive at an expression for the angular momentum of a pseudo-rigid body and an associated equation of balance. In addition, we give generalizations of a number of classical concepts and results, such as body and spatial axes and EULER's classical equations for the motion of a rigid body. Section 4(i) presents a novel view of rigid-body mechanics through the study of rigid motions of a pseudo-rigid body. In particular, we show how Euler angles can be introduced for the study of general pseudo-rigid motions. In Sections 5(iii) and 5(iv), we examine in two different ways the stability of steady spinning motion. The first is an analysis in terms of the components of angular velocity, and direct comparisons are made with classical instabilities exhibited by rigid bodies. The second is an analysis of the stability of a rigid body in terms of Euler angles and ignorable coordinates, following which we explore how the analysis must be generalized to apply to a pseudo-rigid body.

In the second class of comparisons, we look at the theory of pseudo-rigid bodies as a consistent subtheory of nonlinear elasticity. This analysis forms the body of Chapter 3. It includes (1) a discussion of how material frame indifference for pseudo-rigid bodies might derive from this same principle in the three-dimensional theory and (2) a scheme for approximating the mapping, which we have called the coarsifier, that carries pseudo-rigid motions into three-dimensional motions.

The third class of comparisons looks at the theory of pseudo-rigid bodies as a prototype for theories of directed continua. This theme is pervasive in Chapter 2, where the foundations of the subject are developed. In particular, in Section 2(ii), we develop the generalized concept of moment of momentum and give an axiomatic derivation of its constitutive relation. In Section 2(iii), we develop generalized equations

of balance, including one for the Euler tensor, and in Section 2(iv),
we examine in detail generalized loadings for a pseudo-rigid body. In
Section 3(iii), in connection with our comparisons with three-dimensional
theory, we develop interpretations of the generalized measures of strain,
moment of momentum, loads, etc., in terms of more traditional physical
concepts.

The last class of comparisons is developed in Chapter 6, where we
explore ways in which rigid-body mechanics can arise from the theory
of pseudo-rigid bodies in the limit of increasing rigidity. This type of
comparison falls outside the formal theory of consistency as developed
here but nevertheless represents an important means of appraising the
theory. In particular, Section 6(v) is concerned with motions of a pseudo-
rigid body that are small perturbations of the classical motion of a
precessing rigid body.

CHAPTER 2

Foundations of the Theory of Pseudo-rigid Bodies

A pseudo-rigid body, seen from one perspective, is a generalization of a rigid body that accounts for the effects of deformation. In order to lay the foundations for a theory of such bodies, it is useful to begin by revising our view of rigid bodies. While physically we perceive a rigid body to be an object of finite size, analytically we characterize it only through the position of its center of mass and the orientation of its remaining points relative to a set of axes fixed in space. More concisely, a rigid body may be viewed as a moving point to which is attached a time-varying measure of orientation. In this way, we may view the mechanics of rigid bodies as a theory of a special directed continuum. As an immediate generalization, we define a pseudo-rigid body here to be a moving point to which is attached a measure of both orientation and deformation, and we base our formulation of a theory for it on the general principles that have been developed in recent years for the study of directed continua.

(i) Event Worlds and Body Manifolds

At the base of all observations in the physical world is the concept of a space-time manifold $\mathcal{W}$ called the *event world*. The elements w of this manifold are called *events*, and they represent the locations at which we make observations in the real world. In the realm of Newtonian mechanics, the event world is traditionally isomorphic to a product manifold

$$\mathcal{W} \cong \mathcal{E} \times \mathcal{I} \tag{2.1.1}$$

with a spatial component $\mathcal{E}$ called *physical space* and a temporal component $\mathcal{I}$ called *time space*. Tradition also suggests that we model $\mathcal{E}$ as a three-dimensional Euclidean point space and $\mathcal{I}$ as an oriented one-dimensional Euclidean point space. Then any event has a representation $w = (x, i)$ in terms of the *place* x and the *instant* i of that event. We say that the decomposition of the event world in (2.1.1) establishes a *frame of reference*[1] in $\mathcal{W}$.

In practice, we specify events only relative to some fixed event w_0 and often specialize further to the use of coordinates. Let $\mathcal{V}$ and $\mathcal{T}$ be the translation spaces of $\mathcal{E}$ and $\mathcal{I}$, namely, the three- and one-dimensional vector spaces of point differences from $\mathcal{E}$ and $\mathcal{I}$, respectively. If $w_0 = (x_0, i_0)$, then an arbitrary event (x, i) is in one-to-one correspondence with a pair (q, T), $q \in \mathcal{V}$, $T \in \mathcal{T}$, as specified by

$$q = x - x_0, \qquad T = i - i_0. \qquad (2.1.2)$$

We call q and T the *position* and *time* vectors, respectively, of the event (x, i). If we further select an orthonormal basis e_1, e_2, e_3 of $\mathcal{V}$ and a basis I of $\mathcal{T}$, then we can write

$$q = \sum_{i=1}^{3} x_i e_i, \qquad T = tI. \qquad (2.1.3)$$

The scalars x_i are the rectangular Cartesian coordinates of x relative to x_0 and the basis e_i. Similarly, t is the time coordinate of the instant i relative to i_0 and the base vector I. The oriented nature of $\mathcal{I}$ allows us to choose and fix a "past" and a "future." We assume that I maintains this orientation in the sense that if the instant i_2 is after the instant i_1, then the corresponding time coordinates t_1 and t_2 are such that $t_1 < t_2$. Henceforth, following common practice, we assume that a basis I for $\mathcal{T}$ has been selected and fixed, and we work in terms of coordinates t rather than absolute instants i.

In continuum mechanics, a material body is viewed abstractly as (1) a collection of *body points* modeled as a differentiable manifold $\mathcal{B}$, called the *body manifold*, and (2) a class $\mathcal{P}$ of *placements* of this manifold in physical space, generally given by a set of mappings $\chi: \mathcal{B} \to \mathcal{E}$. Intuitively, we see that $\mathcal{P}$ defines the *kinematics* of the theory by specifying the motions, or one-parameter families of placements, that the body can exhibit. By our choice of the body manifold $\mathcal{B}$ and the kinematics prescribed by $\mathcal{P}$, we restrict the type of bodies we choose to consider. For example, through our choice of the dimension of the body manifold $\mathcal{B}$, we can focus on bodies that resemble points, lines, surfaces, or general three-dimensional regions. At the same time, we classify bodies according to those features we choose to observe and measure as we view the body in its motion in the event world $\mathcal{W}$. In the simplest case, we are interested only in the places occupied by the body points at each instant in time. If dim $\mathcal{B} = 0$, we are then led to the mechanics of mass points. Similarly, when dim $\mathcal{B} = 1$, 2, or 3, we are led to studies of strings, membranes, or general three-dimensional continua, respectively.

---

[1] Space-time structures within the context of modern continuum mechanics are discussed extensively by WANG and TRUESDELL [1973], TRUESDELL [1977], and WANG [1979].

Such studies can be specialized further by our choice of the kinematics. For example, if dim $\mathscr{B} = 3$ but in all placements the relative positions of the body points in $\mathscr{E}$ remain unchanged in time, we say that we are studying the mechanics of rigid bodies.[2]

While for many studies in mechanics the foregoing structure of a body is adequate, it requires generalization in order to encompass the broader class of bodies called *directed continua*. Intuitively, we picture such a body once again as a set of body points, but now there is associated with each point some additional structure. Theories of shells and rods are specimens of this picture in which the added structure consists, respectively, of one or two vectors called *directors*.

In order to formalize this picture, we consider the general case of a collection of body points forming a manifold $\mathscr{B}$ of some dimension, and we attach to each point of this manifold a space $\mathscr{D}$ of possibly different dimension. We call this an *extended body* $\mathscr{B}_e$. A natural model for such an object is the *vector bundle*

$$\mathscr{B}_e = \mathscr{B} \times \mathscr{D}. \tag{2.1.4}$$

The base is the body manifold $\mathscr{B}$, and the fiber is a finite-dimensional vector space $\mathscr{D}$, which we call the *director space*. It is this vector space that carries the additional structure of the body.

In order to make observations of this extended body, we must place it in an event world. The event world $\mathscr{W}$ defined in (2.1.1) is not rich enough to permit observations that can discern the new structure that has been added. An *extended event world* $\mathscr{W}_e$ is needed in which we can also follow the behavior in time of the points in the director space $\mathscr{D}$. One choice is

$$\mathscr{W}_e \cong \mathscr{E} \times \mathscr{V} \times \mathscr{I}, \tag{2.1.5}$$

in which the spatial component is now the *tangent bundle* $\mathscr{E} \times \mathscr{V}$ of our conventional physical space $\mathscr{E}$.

Since a directed continuum is pictured as a conventional body with extra structure at each of its points, when we place this extended body in the spatial component $\mathscr{E} \times \mathscr{V}$ we must view its placements as defined by two classes. The first, called the class $\mathscr{P}_b$ of *body placements*, is a set of mappings $\varphi: \mathscr{B} \to \mathscr{E}$ that place the body manifold into physical space. The second, called the class $\mathscr{P}_d$ of *director placements*, is a set of mappings $\Phi: \mathscr{B} \times \mathscr{D} \to \mathscr{V}$, that place the director space $\mathscr{D}$ into the translation space $\mathscr{V}$ of the physical world, possibly with a different

---

[2] If dim $\mathscr{B} = 3$ and all placements of the body are affine transformations of the physical space $\mathscr{E}$, we are led naturally to the approach to pseudo-rigid bodies that was originally proposed by COHEN [1981]. He phrased his work in terms of deforming ellipsoids, but the distinction is small. We adopt here a different approach, taking dim $\mathscr{B} = 0$ and viewing a pseudo-rigid body more naturally, we feel, as a directed continuum.

placement of $\mathscr{D}$ for each of the body points in $\mathscr{B}$. The pair (φ, Φ) is called a *placement* of the extended body $\mathscr{B}_e$ into the event world $\mathscr{W}_e$.

It is not our aim here to develop a general theory of directed continua. We have outlined the preceding structure principally as motivation, and its detail is important only in the case of the following particular body.

Definition. *A* pseudo-rigid body *is an extended body* $\mathscr{B}_e$ *and sets of placements* $\mathscr{P}_b$ *and* $\mathscr{P}_d$ *with the following properties:*

B.I $\mathscr{B}_e$ *is a trivial vector bundle of the form*

$$\mathscr{B}_e = \{P\} \times \mathscr{D} \tag{2.1.6}$$

defined by a single body point P and a three-dimensional director space $\mathscr{D}$.

B.II $\mathscr{B}_e$ *is assigned a positive scalar m, called the* total mass *of the body, and a positive-definite and symmetric bilinear functional M on* $\mathscr{L}(\mathscr{D}) \times \mathscr{L}(\mathscr{D})$, *called the* Euler functional *of the body.* $\mathscr{L}(\mathscr{D})$ *is the dual space of* $\mathscr{D}$.

B.III *The class of body placements is*

$$\mathscr{P}_b = \{\varphi \colon \{P\} \to \mathscr{E}\}, \tag{2.1.7}$$

the set of all mappings of the point P into physical space, and the class of director placements is

$$\mathscr{P}_d = \{\Phi \colon \{P\} \to \mathscr{I}n\nu(\mathscr{D}, \mathscr{V})\}, \tag{2.1.8}$$

the set of all mappings of P into the set of invertible linear maps from $\mathscr{D}$ to $\mathscr{V}$.

The point P represents the center of mass of a continuous three-dimensional body; the space $\mathscr{D}$ is a representation of the matter that surrounds P. The matter in the body is characterized by two measures of mass: (1) the total mass m, and (2) the Euler functional M. The latter is a *generalized measure of mass* associated with the distribution of matter about the center of mass. In order to motivate its definition as a functional, we note that the dual space $\mathscr{L}(\mathscr{D})$ can be viewed as the set of all two-dimensional subspaces of the vector space $\mathscr{D}$. Physically, we interpret this as the set of all planes in the physical space $\mathscr{E}$ passing through the point P. The more abstract interpretation of elements N of $\mathscr{L}(\mathscr{D})$ as linear functionals on $\mathscr{D}$ is included here by viewing $N(V)$ as the "distance" of V from the plane N through P. In Euclidean space, this distance is simply the usual orthogonal projection of V onto the unit normal to N, and it is customary to highlight this interpretation by introducing the notation

$$N(V) = N \cdot V, \qquad V \in \mathscr{D}. \qquad (2.1.9)$$

This form brings out the usual dual character[3] of $\mathscr{D}$ and $\mathscr{L}(\mathscr{D})$. For the Euler functional M, we interpret the value $M(N, K)$ as the moment, weighted by the distribution of mass, of the product of the distances to the two planes N and K through the point P. Thus, M can be viewed as a second moment of "distance" relative to the center of mass, arising through some weighted integration over $\mathscr{D}$. Since P lies at the center of mass, necessarily the associated first moments vanish identically.

The central component of this definition is the class of placements prescribed by **B.III**. Each placement is a pair of mappings $\lambda = (\varphi_\lambda, \Phi_\lambda)$, and its effect can be represented explicitly by the equations

$$x = \varphi_\lambda(P),$$
$$v = \Phi_\lambda(P)V, \qquad V \in \mathscr{D}, \qquad (2.1.10)$$

where $\varphi_\lambda(P) \in \mathscr{E}$ and $\Phi_\lambda(P) \in \mathscr{I}nv(\mathscr{D}, \mathscr{V})$. The place $\varphi_\lambda(P)$ is the location in $\mathscr{E}$ of the center of mass of the body. Moreover, the mapping $(2.1.10)_2$ shows that the matter surrounding P, as located by vectors V in $\mathscr{D}$ or vectors v in $\mathscr{V}$, is important only with respect to changes produced by an invertible linear transformation. In essence, the linear transformation is the new structure that has been added to our point manifold. Equivalently, our body is a single point to which are attached three vectors, called *directors*, that form a basis for the vector space $\mathscr{D}$. We explore this interpretation in more detail later.

It is useful to mention some other directed bodies that arise from slight modifications of the preceding definition. If we replace $\mathscr{I}nv(\mathscr{D}, \mathscr{V})$ with the set of all *isometries* from $\mathscr{D}$ to $\mathscr{V}$, we obtain a point body to which is assigned a measure of orientation. This is one way to define a *rigid body*. Next, let us replace $\{P\}$ by an n-dimensional body manifold $\mathscr{B}$, where $1 \le n \le 2$, and let $\mathscr{D}$ be a $(3 - n)$-dimensional vector space. Let $\mathscr{T}(P)$ be the tangent space of $\mathscr{B}$ at the point P. Define $\mathscr{P}_b$ to be the set of all smooth mappings φ of $\mathscr{B}$ into $\mathscr{E}$, with φ^* denoting the *derivative map* of φ. Then let $\mathscr{P}_d$ be the set of all mappings Φ of $\mathscr{B}$ into the space of linear mappings from $\mathscr{D}$ to $\mathscr{V}$ such that, for each $P \in \mathscr{B}$, the local transformation $(\varphi^*(P), \Phi(P))$ lies in $\mathscr{I}nv(\mathscr{T}(P) \times \mathscr{D}, \mathscr{V})$. In this way, we obtain an n-dimensional body at each point of which is attached a set of $3 - n$

---

[3] The representation of $\mathscr{L}(\mathscr{D})$ as a set of planes is discussed by SCHOUTEN [1954]. Indeed, his formulation includes for any plane the assignment of its orientation in $\mathscr{D}$. Thus, the "distance" defined by (2.1.9) is often called the *scalar product*. Since the vector spaces we consider here are generally finite dimensional, we may identify in the standard way a vector space and its dual. Under this identification, the scalar product reduces to the standard *inner product* in Euclidean space, and elements of $\mathscr{L}(\mathscr{D})$ can be viewed equivalently as unit normals to the planes in $\mathscr{E}$. Here, however, it is useful to maintain the distinction between a space and its dual, and so (2.1.9) should be viewed simply as a short-hand notation for the action of an element of a dual space on some element of the parent space.

directors. When $n = 1$, this is a *rod*, and if $n = 2$, we obtain a *shell*. Moreover, by restricting attention to special subsets of $\mathscr{P}_d$, we can obtain rods with rigid cross sections, shells of fixed thickness, and a myriad of other possibilities.

A *pseudo-rigid motion* is a smooth[4] mapping $t \mapsto \chi(t) = (\varphi_{\chi(t)}, \Phi_{\chi(t)})$ of an interval of time $\mathscr{I} \subset \mathscr{T}$ into the set of placements of $\mathscr{B}_e$. It can be represented by the equations

$$x = \varphi_{\chi(t)}(P) \equiv \varphi(P, t),$$
$$v = \Phi_{\chi(t)}(P)V \equiv \Phi(P, t)V, \qquad V \in \mathscr{D}. \tag{2.1.11}$$

At each time t the point P occupies the place $\varphi(P, t)$ in $\mathscr{E}$, which we interpret as the current location of the center of mass. The transformation $\Phi(P, t)$ provides a representation of how the matter surrounding P has deformed at time t.

Through (2.1.10) we have transferred the body point P and its neighboring matter into physical space. It is useful at this stage to transfer also the two measures of mass. Generally, the total mass associated with the point $\varphi_\lambda(P)$ is a number m_λ. For simplicity, we assign the total mass m of the body directly to this point. Thus, we require that

$$m_\lambda = m \tag{2.1.12}$$

for every placement λ. In particular, in a pseudo-rigid motion the mass $m_{\chi(t)}$ is the constant m for all t, and so we explicitly require that the total mass of a body remains constant in time:

$$\frac{dm}{dt} = 0. \tag{2.1.13}$$

Here and henceforth we write m in place of $m_{\chi(t)}$.

As a preliminary to similar considerations for the Euler functional M, let us note that associated with any linear mapping $\Phi_\lambda(P)$ from $\mathscr{D}$ to $\mathscr{V}$ there is a *dual mapping* $\Phi_\lambda^\mathsf{T}(P)$ from $\mathscr{L}(\mathscr{V})$ to $\mathscr{L}(\mathscr{D})$. Setting

$$N = \Phi_\lambda^\mathsf{T}(P)n, \qquad n \in \mathscr{L}(\mathscr{V}),$$
$$n = (\Phi_\lambda^\mathsf{T}(P))^{-1}N, \qquad N \in \mathscr{L}(\mathscr{D}), \tag{2.1.14}$$

we may interpret n as an image plane in the placement λ corresponding to the plane N, or view N as the preimage plane associated with the plane n in λ. In order to obtain a clear physical interpretation of the image normal n, we recall the definition of the dual mapping. For any linear mapping A from $\mathscr{D}$ to $\mathscr{V}$, its dual A^T is that linear transformation from $\mathscr{L}(\mathscr{V})$ to $\mathscr{L}(\mathscr{D})$ such that for any $V \in \mathscr{D}$ and any $n \in \mathscr{L}(\mathscr{V})$ we have $n \cdot (AV) = (A^\mathsf{T}n) \cdot V$. In view of (2.1.14), this implies that

$$N \cdot V = (\Phi_\lambda^\mathsf{T}(P)n) \cdot V = n \cdot \Phi_\lambda(P)V. \tag{2.1.15}$$

---

[4] Nominally, we require two derivatives, but generally in this monograph we let the analysis dictate the degree of regularity.

Since $v = \mathbf{\Phi}_\lambda(P)V$, we obtain

$$N \cdot V = n \cdot v. \tag{2.1.16}$$

Therefore, in moving from the extended body $\mathscr{B}_e$ to physical space, we find that "distances" from planes, as measured by the action of linear functionals, remain unchanged. The central notion here is the invariance of "distance" as summarized in (2.1.16).

In carrying the Euler functional M to physical space, we generalize the preceding observation by viewing the values $M(N, K)$ as invariant. Thus, for each placement λ we introduce an Euler functional E_λ on physical space by requiring that

$$M(N, K) = E_\lambda(n, k) \tag{2.1.17}$$

for every choice of the planes n and k in $\mathscr{L}(\mathscr{V})$ related, respectively, to the planes N and K in $\mathscr{L}(\mathscr{D})$ according to (2.1.14). This relation uniquely defines E_λ. Naturally, each plane represented here by an element of $\mathscr{L}(\mathscr{V})$ is assumed to pass through the center of mass of the body in the placement λ, namely, through $\varphi_\lambda(P)$.

In a pseudo-rigid motion, this same process defines an Euler functional $E_{\chi(t)}$ for each time t. Henceforth, we drop the subscript $\chi(t)$ and simply write $E(t)$. While E may depend explicitly on t, *we assume that the Euler functional M on the extended body $\mathscr{B}_e$ is constant.* If we also view the planes N and K in $\mathscr{L}(\mathscr{D})$ to be fixed, then their images in $\mathscr{L}(\mathscr{V})$, namely,

$$n(t) = \mathbf{\Phi}^{-\mathrm{T}}(P, t)N, \qquad k(t) = \mathbf{\Phi}^{-\mathrm{T}}(P, t)K, \tag{2.1.18}$$

will generally change in time. (The superscript $-\mathrm{T}$ denotes the dual of the inverse map, or equivalently the inverse of the dual map.) However, these changes will occur, according to (2.1.17), such that $E(n, k)$ always has the constant value $M(N, K)$. Thus,

$$\frac{d}{dt} E(n, k) = 0, \tag{2.1.19}$$

paralleling (2.1.13) for the total mass.

We can interpret (2.1.13) and (2.1.19) as *conservation laws* for the total mass and the Euler functional. They arise directly out of our definitions of m and E in physical space and our assumptions that m, M, and elements of $\mathscr{L}(\mathscr{D})$ on the extended body $\mathscr{B}_e$ are constant. While there are many interesting problems in mass-point mechanics that allow for changes in the mass through time, such problems are seldom considered in the mechanics of rigid bodies and are even more rare in modern continuum mechanics.[5]

---

[5] Indeed, the mass density ρ for a continuum is conventionally required to satisfy an equation of continuity, and this equation can be traced directly to the assumption that the total mass be invariant from one placement of the body to another. Cf. TRUESDELL [1977, §II.2].

In the mechanics of rigid bodies, it is conventional to introduce an Euler tensor rather than an Euler functional. Since E is a bilinear functional on $\mathscr{L}(\mathscr{V})$, it can be represented in the form

$$E(\boldsymbol{n}, \boldsymbol{k}) = \boldsymbol{n} \cdot \boldsymbol{E}\boldsymbol{k}, \qquad (2.1.20)$$

and we call $\boldsymbol{E} \colon \mathscr{L}(\mathscr{V}) \to \mathscr{V}$ the *Euler tensor* of the motion. For theoretical developments, the Euler functional plays a principal role, but, as we shall see later, the Euler tensor makes an explicit appearance when we consider specific applications of the theory. For comparisons with results in rigid-body mechanics, it is useful to introduce the *inertia tensor* in a pseudo-rigid motion by

$$\boldsymbol{J} = (\operatorname{tr} \boldsymbol{E})\boldsymbol{I} - \boldsymbol{E}. \qquad (2.1.21)$$

Naturally, by replacing E, $\boldsymbol{E}$, and $\boldsymbol{J}$ here with E_λ, $\boldsymbol{E}_\lambda$, and $\boldsymbol{J}_\lambda$, we may define an Euler tensor and an inertia tensor for any placement λ of the body.

(ii) Kinematics

In continuum mechanics, some particular placement $\kappa = (\varphi_R, \Phi_R)$ is often used as an absolute reference relative to which we may compare the placements of the body during its motion. We call it a *reference placement*. Because of its special status, we denote it by the subscript R and generally we use this subscript to denote all quantities relative to such a placement. Let

$$\begin{aligned} X &= \varphi_R(P), \\ V &= \Phi_R(P)V, \qquad V \in \mathscr{D}, \end{aligned} \qquad (2.2.1)$$

where we recall that $\Phi_R(P) \in \mathscr{I}nv(\mathscr{D}, \mathscr{V})$. Then X and V denote independent variables representing places and vectors, respectively, in the reference placement. In view of $(2.1.2)_1$ we see that the motion of a pseudo-rigid body, generally given by $(2.1.11)$, can now be represented equivalently in the form

$$\begin{aligned} \boldsymbol{q} &= \boldsymbol{r}(t), \\ \boldsymbol{v} &= \boldsymbol{F}(t)V, \qquad V \in \mathscr{V}, \end{aligned} \qquad (2.2.2)$$

where

$$\boldsymbol{r}(t) = \boldsymbol{\varphi}(t) - \boldsymbol{x}_0, \qquad \boldsymbol{F}(t) = \boldsymbol{\Phi}(t)\boldsymbol{\Phi}_R^{-1}. \qquad (2.2.3)$$

Here and henceforth we suppress dependence on the single material point P. The vector $\boldsymbol{r}$ is the *displacement* of the mass center of the body relative to the origin $\boldsymbol{x}_0$, and $\boldsymbol{F}$ is the *deformation* of the body relative to κ. The pair $(\boldsymbol{r}, \boldsymbol{F})$ is called the *transplacement* relative to κ, and it provides a representation of the motion of a pseudo-rigid body through a *referential description*. Note that, in this description, $F(t)$ is any member of the general linear group $\mathcal{Gl}(\mathcal{V})$ on $\mathcal{V}$.

For a rigid body, whose director placements analogous to (2.1.8) must be isometries from $\mathcal{D}$ to $\mathcal{V}$, the referential description (2.2.2) still applies, except that $F(t)$ must lie in the orthogonal group $\mathcal{O}(\mathcal{V})$ on $\mathcal{V}$.

The *velocity* s and *acceleration* a of the center of mass of a pseudo-rigid body are represented by the usual time derivatives

$$s = \dot{r}, \qquad a = \ddot{r}. \tag{2.2.4}$$

Velocities and accelerations of neighboring points are modeled by successive derivatives of v, holding its preimage V constant under transplacement. These derivatives, expressed relative to v itself, are given by

$$\dot{v} = Lv, \qquad \ddot{v} = (\dot{L} + L^2)v, \tag{2.2.5}$$

where

$$L = \dot{F}F^{-1} \tag{2.2.6}$$

is the *deforming tensor*. Thus, L and $\dot{L} + L^2$ arise naturally as representations of the velocity and acceleration associated with the deformation F. The skew part of L, denoted by sk L, is the *spin tensor* O of the pseudo-rigid motion, while the symmetric part sym L is the *stretching tensor* D. Thus,

$$L = D + O, \tag{2.2.7}$$

where

$$\begin{aligned} D &= \text{sym } L = \tfrac{1}{2}(L + L^\mathsf{T}), \\ O &= \text{sk } L = \tfrac{1}{2}(L - L^\mathsf{T}). \end{aligned} \tag{2.2.8}$$

If F is written in terms of its polar decomposition

$$F = RU, \tag{2.2.9}$$

where R is orthogonal and U is positive-definite and symmetric, then the stretching and spin tensors are given by the conventional expressions (cf. TRUESDELL and NOLL [1965, Eq. (24.16)])

$$\begin{aligned} D &= \tfrac{1}{2}R(\dot{U}U^{-1} + U^{-1}\dot{U})R^\mathsf{T}, \\ O &= \dot{R}R^\mathsf{T} + \tfrac{1}{2}R(\dot{U}U^{-1} - U^{-1}\dot{U})R^\mathsf{T}. \end{aligned} \tag{2.2.10}$$

The theory of pseudo-rigid bodies provides an important tool for studying the interplay between effects of orientation, as characterized by R, and effects of pure deformation, as represented by U. The tensor $\dot{R}R^\mathsf{T}$, which is skew, is called the *angular velocity tensor* W of the motion. Its associated axial vector is called the *angular velocity vector* w. It is interesting to note that the spin tensor O provides an alternative to the angular velocity tensor for examining effects of rotation. Generally, we will find that the latter is more convenient for applications. While the spin and the angular velocity are generally distinct measures of rotation, for a rigid motion the pure deformation U is trivially I, and in this case it is clear from $(2.2.10)_2$ that the spin and the angular velocity agree.

With the placement κ we can associate a *referential Euler functional* E_R and a *referential Euler tensor* E_R. According to the prescriptions (2.1.17) and (2.1.20), we set

$$M(N, K) = E_R(N, K) = N \cdot E_R K, \qquad (2.2.11)$$

where

$$N = \Phi_R^{-T} N, \qquad K = \Phi_R^{-T} K. \qquad (2.2.12)$$

In this case, the planes N and K pass through the center of mass φ_R in the reference placement. Since the *current Euler functional* E was defined so that in any pseudo-rigid motion its value $E(n, k)$, or equivalently $n \cdot Ek$, agrees with $M(N, K)$, we find that

$$N \cdot E_R K = n \cdot Ek. \qquad (2.2.13)$$

By combining (2.2.12) and (2.1.18) and using the definition $(2.2.3)_2$ of the deformation F, we find that the current planes n and k are related to their companions N and K in the reference placement by

$$n = F^{-T}N, \qquad k = F^{-T}K. \qquad (2.2.14)$$

Of course n and k pass through the current center of mass φ. We see now that

$$n \cdot Ek = F^T n \cdot E_R F^T k = n \cdot F E_R F^T k, \qquad (2.2.15)$$

and since n and k are arbitrary, we obtain

$$E(t) = F(t) E_R F^T(t). \qquad (2.2.16)$$

The conservation of the Euler functional summarized in (2.1.19) was based on the assumption that on the body manifold $\mathcal{B}_e$ the Euler functional M and the elements N and K of $\mathcal{L}(\mathcal{D})$ are independent of time. Since by nature the reference placement κ is independent of t, we conclude from (2.2.11) that the referential Euler tensor E_R is also constant. Thus, (2.2.16) shows that the true dependence of E on time arises through the deformation F. In essence, (2.2.16) provides an integral version[6] of the conservation principle (2.1.19), or equivalently

$$\dot{E}_R = 0. \qquad (2.2.17)$$

---

[6] This result has an exact parallel from continuum mechanics. Conventionally, the mass density ρ in the current placement satisfies the equation of continuity $\partial_t \rho + \operatorname{div} \rho u = 0$, with u denoting the field of velocity. This equation parallels (2.1.19). Equivalently, we can write $\partial_t \rho_R = 0$, paralleling (2.2.17), where ρ_R and ρ are related by the integral relation $\rho \det F = \rho_R$. This last relation is the continuum analogue of (2.2.16).

EULER'S laws of balance, and the generalizations of these on which we shall base the governing equations for the theory of pseudo-rigid bodies, assert that external effects cause changes in measures of the momentum of a body. Our choices of these measures of momentum reflect our intuition about what effects cause a body to persist in its motion when external influences are absent. Generally, we build this intuition either by generalizing from simpler theories or simplifying from the more complex. The motion of a mass point is characterized by a curve in space-time, and the momentum of this mass is a vector proportional to both its mass and its velocity. In the absence of external forces, increases in mass result in decreases in velocity, exactly in a proportion to keep the momentum fixed. For rigid bodies, whose motions also involve changes in orientation, we consider in addition the effects of the angular momentum of a body. In this case, the rate of spin combines with moments of inertia in order to hold the angular momentum vector fixed.

For a pseudo-rigid body $\mathscr{B}_e$, we generalize these ideas by characterizing momentum through two quantities, a vector $\boldsymbol{p} \in \mathscr{V}$ called the *linear momentum* and a linear mapping $\boldsymbol{h}_R \colon \mathscr{L}(\mathscr{V}) \to \mathscr{V}$ called the *referential moment of momentum*. Within the general framework of directed continua, the latter quantity might also be called the *generalized momentum* of the body. For each plane $N \in \mathscr{L}(\mathscr{V})$, we interpret $\boldsymbol{h}_R(N)$ as the net moment, with respect to *distance* from N in the reference placement, of the linear momenta of all points in the body. Since $\boldsymbol{h}_R$ is a linear mapping, we can represent it in the form

$$\boldsymbol{h}_R(N) = \boldsymbol{H}_R N, \tag{2.2.18}$$

where $\boldsymbol{H}_R$ is called the *referential moment of momentum tensor*. As we shall see later, $\boldsymbol{H}_R$ contains the conventional effects of angular momentum as well as new effects that are particular to pseudo-rigid bodies.

In the same way, a moment of momentum can be introduced in connection with any placement of the body. In particular, if we work relative to the current placement $\chi(t)$ in a pseudo-rigid motion, we obtain the *current moment of momentum* $\boldsymbol{h} \colon \mathscr{L}(\mathscr{V}) \to \mathscr{V}$; the associated *current moment of momentum tensor* $\boldsymbol{H}$ is given by

$$\boldsymbol{h}(n) = \boldsymbol{H}n. \tag{2.2.19}$$

We can interrelate the moments of momentum for different placements by requiring that their values be invariant. For example, the referential and current values must satisfy

$$\boldsymbol{h}(n) = \boldsymbol{h}_R(N) \tag{2.2.20}$$

for all n and N related through (2.2.14)$_1$. By (2.2.18) and (2.2.19) we conclude that

$$\boldsymbol{H} = \boldsymbol{H}_R \boldsymbol{F}^T(t). \tag{2.2.21}$$

Experience drawn from the study of mass points, rigid bodies, and general three-dimensional continua suggests that p and H_R should be determined by the motion of the body through measures of mass and velocity. Thus, constitutive relations must be specified for each, giving their values in terms of m, E_R, and the transplacement (r, F) relative to the reference placement κ. For the linear momentum p we choose the standard form from rigid-body mechanics:

$$p = m\dot{r}. \tag{2.2.22}$$

For the moment of momentum H_R, however, a corresponding choice is less clear. Since H_R principally captures the effects of the linear momentum of points surrounding the center of mass, and since mass and velocity for such points are characterized, respectively, by the Euler tensor E_R and the rate of change $\dot{F}$ of the deformation, it is reasonable to begin with a general relation of the form

$$H_R = \mathfrak{H}_R(\dot{F}, E_R). \tag{2.2.23}$$

In order to simplify the response function $\mathfrak{H}_R$, we adopt two additional requirements suggested by the physical setting.

First, we note that, without further restrictions on (2.2.23), $\mathfrak{H}_R$ will change with our choice of the reference placement κ. Thus, different placements will generally see different ways of calculating the moment of momentum tensor of the body. It seems more reasonable, however, to assume that the calculation of the moment of momentum is not connected to any particular choice of the reference placement. Thus, we adopt the following:

(H1) *The response function $\mathfrak{H}_R$ for the moment of momentum is invariant under changes of reference placement.*

More precisely, let $\hat{\kappa} = (\varphi_S, \Phi_S)$ be a second choice for the reference placement, with Euler tensor E_S, moment of momentum tensor H_S, and response function $\mathfrak{H}_S$. If G denotes the deformation relative to $\hat{\kappa}$, namely, $G = \Phi(t)\Phi_S^{-1}$, then the constitutive relation (2.2.23) relative to $\hat{\kappa}$ becomes

$$H_S = \mathfrak{H}_S(\dot{G}, E_S). \tag{2.2.24}$$

(H1) then states that

$$\mathfrak{H}_S = \mathfrak{H}_R \tag{2.2.25}$$

for all placements $\hat{\kappa}$.

If a pseudo-rigid body in a particular reference placement has a completely symmetric distribution of mass, so that E_R is proportional to I, then it is reasonable to view all points in the body as independent mass points with moment of momentum proportional to $\dot{F}$. This motivates our second requirement:

(H2) *For all scalars ξ and all tensors K,*

$$\mathfrak{H}_R(K, \xi I) = \xi K. \tag{2.2.26}$$

These two hypotheses imply a unique form for the response function for moment of momentum:

Theorem. *The only function $\mathfrak{H}_R$ satisfying both (H1) and (H2) is*

$$\mathfrak{H}_R(\dot{F}, E_R) = \dot{F} E_R. \tag{2.2.27}$$

Thus, by (2.2.21), (2.2.23), (2.2.6), and (2.2.16), the constitutive relations for the referential and current moment of momentum tensors are

$$H_R = \dot{F} E_R, \qquad H = LE. \tag{2.2.28}$$

Later, we shall see that the skew part of the tensor H provides an exact analogue of the conventional angular momentum from rigid-body mechanics.

We conclude this section with the proof of (2.2.27). Let P denote the transformation $\Phi_S \Phi_R^{-1}$ from the placement κ to the placement $\hat{\kappa}$. Since $F = \Phi \Phi_R^{-1}$ and $G = \Phi \Phi_S^{-1}$, we see that

$$F = GP. \tag{2.2.29}$$

Moreover, (2.2.16) and (2.2.21), specialized now to the placement $\hat{\kappa}$ rather than the current placement, give us

$$E_S = P E_R P^T, \qquad H_S = H_R P^T. \tag{2.2.30}$$

Placing these into (2.2.24) and using (2.2.25) and (2.2.23), we obtain

$$\mathfrak{H}_R(\dot{G}, P E_R P^T) = \mathfrak{H}_R(\dot{G} P, E_R) P^T. \tag{2.2.31}$$

Setting $E_R = I$, we see from (H2) that

$$\mathfrak{H}_R(\dot{G}, P P^T) = \mathfrak{H}_R(\dot{G} P, I) P^T = \dot{G} P P^T. \tag{2.2.32}$$

Since this applies for all invertible tensors P, choose P to be the square root of the positive-definite and symmetric tensor E_R. Then, replacing G with F, we obtain (2.2.27).

(iii) Equations of Balance

In rigid-body mechanics, the dynamic rules that govern how a body moves in physical space express principles of balance between (1) changes in kinematical features of the body and (2) external influences that cause motion. These principles are the laws of balance of *linear* and *angular momentum* due to EULER. To these we may adjoin the (seldom stated) principles of balance of mass and the Euler functional. Classically, external influences are modeled by the *external force* and the *external torque* on the body. Since the theory of pseudo-rigid bodies is a generalization of rigid-body mechanics, we may expect that dynamical principles for it will generalize in a simple way the classical laws and that the external influences on a pseudo-rigid body will represent simple generalizations of external forces and torques.

We assume that pseudo-rigid bodies move in response to two agents, the *external force*, as modeled by a vector $f \in \mathcal{V}$, and the *referential external force-moment*, as modeled by a linear mapping $m_R: \mathcal{L}(\mathcal{V}) \to \mathcal{V}$. The former is precisely that quantity which appears in rigid-body mechanics. It represents the net force on the body owing to those concentrated loads, loads distributed as fields over the body, and surface tractions that may be acting. The concept of an external force-moment is introduced more conventionally within the context of theories of generalized continua. The value $m_R(N)$ represents the net moment, with respect to distance from the referential plane $N \in \mathcal{L}(\mathcal{V})$, of all forces that are acting on the body.

Continuum mechanics views the concept of *force* as primitive, without definition in terms of other quantities. Force is something we sense in nature, through pressure, pulls, and pushes, and we model it mathematically in various ways according to our intuition and experience about the way it acts in a given situation. In particular, force is assumed to be an additive quantity because this is our perception of how its effects cumulate. In keeping with these ideas, we assume that the external force f and the external force-moment m_R are *additive primitive quantities*. From the point of view of the foundations· of our subject, this is a natural assumption. However, it leaves open an important question that arises as soon as we wish to examine specific problems; namely, how do we model the force and force-moment mathematically in each problem we confront? We address this issue in some depth in the next section.

The motion of a pseudo-rigid body is also affected by internal forces. These are represented by a quantity σ_R of the same type as m_R and called the *referential internal force-moment*. Physically, $\sigma_R(N)$ measures the resultant of the moments of the internal forces with respect to distance from the plane N in the reference placement. σ_R is intimately related to the deformation of $\mathcal{B}_e$, and hence its dependence on our measures of deformation must be prescribed by a constitutive relation. We assume this relation takes the form

$$\sigma_R(N) = s_R(F^t, N), \qquad (2.3.1)$$

where s_R is called the *response functional* of the pseudo-rigid body. The notation indicates that s_R is generally a function of the *history* F^t of the deformation, F^t denoting the set of values $F(s)$ for all past times $s \leq t$. This assumption, if we may borrow from the terminology of modern continuum mechanics, defines a pseudo-rigid body with *simple material behavior*. The response functional also carries information about the state of the body in the reference placement. Special material symmetries and internal loadings are examples of this type of detail. Generally, such features depend on our choice of placement, and so s_R is also a function of the reference placement κ. For notational convenience we suppress this dependence throughout our development.

The governing dynamical principles of the theory of pseudo-rigid bodies are direct generalizations of EULER's laws. We state them in the referential form[7]

E.I $\quad \dfrac{d}{dt} m = 0,$

E.II $\quad \dfrac{d}{dt} E_R(N, K) = 0,$

$$(2.3.2)$$

E.III $\quad \dfrac{d}{dt} p = f,$

E.IV $\quad \dfrac{d}{dt} h_R(N) = m_R(N) - \sigma_R(N).$

These are the *equations of balance*. The first two are the *principles of balance of mass and the Euler functional*, respectively. As indicated earlier, together they state that the mass distribution in space, as viewed from the reference placement, is independent of time. The third equation is the *principle of balance of linear momentum* common to both mass-point and rigid-body mechanics. The final equation is a generalization of EULER's balance of angular momentum called the *principle of balance of moment of momentum*. As we shall see in Section 2(v), it contains in part exactly EULER's conventional assumption.

For certain applications, as well as for general comparisons with rigid-body mechanics, it is useful to restate our generalizations of EULER's laws in terms of the current placement of the body. In order to do this, we must first carry the external and internal force-moments m_R and σ_R to the current placement. In analogy with the moment of momentum h_R, we assume that $m_R(N)$ and $\sigma_R(N)$ are invariant with respect to changes of placement. In this way we may introduce the *current external force-moment* m and the *current internal force-moment* σ by the invariance conditions

$$m(n) = m_R(N), \qquad \sigma(n) = \sigma_R(N). \qquad (2.3.3)$$

Naturally, σ is given by a corresponding constitutive relation of the form

$$\sigma(n) = s(F^t, n). \qquad (2.3.4)$$

---

[7] These are stated with respect to an *inertial* frame of reference.

By (2.2.13), (2.2.20), and (2.3.3), we see that the generalization (2.3.2) of EULER'S laws can now be written in the spatial form

$$\frac{d}{dt}m = 0,$$

$$\frac{d}{dt}E(\boldsymbol{n}, \boldsymbol{k}) = 0,$$

$$\frac{d}{dt}\boldsymbol{p} = \boldsymbol{f},$$

(2.3.5)

$$\frac{d}{dt}\boldsymbol{h}(\boldsymbol{n}) = \boldsymbol{m}(\boldsymbol{n}) - \boldsymbol{\sigma}(\boldsymbol{n}).$$

When combined with the constitutive relations (2.2.22), (2.2.23), (2.3.1), and (2.3.4), the preceding equations of balance give rise to equations of motion, namely, explicit differential equations for $\boldsymbol{r}$ and $\boldsymbol{F}$ (and, of course, m and E). We now derive these equations. A novel feature of the calculation is the fact that the planes $\boldsymbol{n}$, $\boldsymbol{k}$, $\boldsymbol{N}$, and $\boldsymbol{K}$ appear explicitly in our generalizations of EULER'S laws, and we must eliminate them before we can obtain equations for $\boldsymbol{r}$ and $\boldsymbol{F}$ alone. One step in this direction is straightforward. Since $\boldsymbol{m}_{\mathrm{R}}$ and $\boldsymbol{\sigma}_{\mathrm{R}}$ are linear functions on $\mathscr{L}(\mathscr{V})$, we can represent them in terms of tensors. Thus,

$$\boldsymbol{m}_{\mathrm{R}}(\boldsymbol{N}) = \boldsymbol{M}_{\mathrm{R}}\boldsymbol{N}, \qquad \boldsymbol{\sigma}_{\mathrm{R}}(\boldsymbol{N}) = \boldsymbol{\Sigma}_{\mathrm{R}}\boldsymbol{N} = \mathfrak{S}_{\mathrm{R}}(\boldsymbol{F}^t)\boldsymbol{N}, \qquad (2.3.6)$$

where we also refer to $\boldsymbol{M}_{\mathrm{R}}$ and $\boldsymbol{\Sigma}_{\mathrm{R}}$ as the *referential external* and *internal force-moments*, respectively. The response function $\mathfrak{S}_{\mathrm{R}}$ expresses $\boldsymbol{\Sigma}_{\mathrm{R}}$ in terms of the history $\boldsymbol{F}^t$ of the deformation, and, in analogy with $\mathfrak{s}_{\mathrm{R}}$, we suppress the dependence of $\mathfrak{S}_{\mathrm{R}}$ on the reference placement κ. The *current external force-moment* $\boldsymbol{M}$ and the *current internal force-moment* $\boldsymbol{\Sigma}$ and its response function $\mathfrak{S}$ are defined similarly:

$$\boldsymbol{m}(\boldsymbol{n}) = \boldsymbol{M}\boldsymbol{n}, \qquad \boldsymbol{\sigma}(\boldsymbol{n}) = \boldsymbol{\Sigma}\boldsymbol{n} = \mathfrak{S}(\boldsymbol{F}^t)\boldsymbol{n}. \qquad (2.3.7)$$

By calculations paralleling those leading to (2.2.21), we find that the referential and current forms of these tensors are related by

$$\boldsymbol{M} = \boldsymbol{M}_{\mathrm{R}}\boldsymbol{F}^{\mathrm{T}}, \qquad \boldsymbol{\Sigma} = \boldsymbol{\Sigma}_{\mathrm{R}}\boldsymbol{F}^{\mathrm{T}}, \qquad \mathfrak{S}(\boldsymbol{F}^t) = \mathfrak{S}_{\mathrm{R}}(\boldsymbol{F}^t)\boldsymbol{F}^{\mathrm{T}}. \qquad (2.3.8)$$

Let us turn now to a closer examination of the planes $\boldsymbol{n}$, $\boldsymbol{k}$, $\boldsymbol{N}$, and $\boldsymbol{K}$. Since $\boldsymbol{n}$ and $\boldsymbol{k}$ are the images of $\boldsymbol{N}$ and $\boldsymbol{K}$ under transplacement, they are defined explicitly by (2.2.14). This equation shows that we may take as independent variables either $\boldsymbol{N}$ and $\boldsymbol{K}$ or $\boldsymbol{n}$ and $\boldsymbol{k}$, but not both pairs simultaneously. If we fix $\boldsymbol{N}$ and $\boldsymbol{K}$ in time, (2.2.14) implies a specific time dependence for $\boldsymbol{n}$ and $\boldsymbol{k}$, and this must be considered when we compute the time derivatives appearing in (2.3.5)$_{2,4}$. Alternately, we could fix $\boldsymbol{n}$ and $\boldsymbol{k}$, or more generally give one of these pairs some specific variation in time and let (2.2.14) determine the subsequent variation of the other pair. While the theory does not restrict this choice, tradition suggests that independent variables in the reference placement be viewed as independent of time. For this reason we consider N and K henceforth as constants, with n and k varying with the deformation F according to (2.2.14). Physically, this means that all moments of inertia, moments of momentum, and force-moments in the reference placement are computed relative to fixed planes in the body.

Consider now the referential form (2.3.2) of EULER's laws. Substituting the representations $(2.2.11)_2$, (2.2.22), (2.2.18), $(2.2.28)_1$, and (2.3.6) and using the fact that N and K are independent of time and also arbitrary, we obtain the following *equations of motion in referential form*:

$$\dot{m} = 0, \qquad \dot{E}_R = 0,$$
$$m\ddot{r} = f, \qquad \dot{F}E_R = M_R - \mathfrak{S}_R(F').$$

(2.3.9)

Chapters 4, 5, and 6 are devoted to the analysis of solutions of these equations, both in general and in applications to specific problems.

Consider next the spatial equations of balance (2.3.5). By (2.2.14), (2.2.6), and the identity $(F^{-1})' = -F^{-1}\dot{F}F^{-1}$, we find that

$$\dot{n} = -F^{-T}\dot{F}^T F^{-T} N = -(\dot{F}F^{-1})^T n = -L^T n, \qquad (2.3.10)$$

and similarly

$$\dot{k} = -L^T k. \qquad (2.3.11)$$

Thus,

$$\frac{d}{dt}E(n, k) = \frac{d}{dt}n \cdot Ek$$
$$= \dot{n} \cdot Ek + n \cdot \dot{E}k + n \cdot E\dot{k}$$
$$= n \cdot (\dot{E} - EL^T - LE)k,$$
$$\frac{d}{dt}h(n) = \frac{d}{dt}Hn$$
$$= \dot{H}n + H\dot{n}$$
$$= (\dot{H} - HL^T)n.$$

(2.3.12)

Therefore, by (2.3.7) and the fact that n and k can now be viewed as arbitrary, we obtain the *spatial equations of motion*

$$\dot{m} = 0, \qquad \dot{E} - EL^T - LE = 0,$$
$$m\ddot{r} = f, \qquad (LE)' - LEL^T = M - \mathfrak{S}(F').$$

(2.3.13)

The differential equation here for E appears to complicate the theory considerably. Fortunately, it has an exact solution, namely, that given in (2.2.16) where E_R is viewed as a constant of integration. Nevertheless, the very fact that E is an explicit function of time makes these spatial equations less attractive for applications than the referential forms (2.3.9).

The dependence of n and k on time has led to the terms $-EL^T$, $-LE$, and $-LEL^T$ on the left-hand sides of $(2.3.13)_{2,4}$. As a result these equations bear little resemblance to conventional spatial equations of balance in continuum mechanics. While this is not of particular concern for subsequent applications, a close analogy with other continuum theories is useful in regard to the foundations of our subject. The following is one way to achieve this analogy.

The over dot we used previously denotes the conventional ordinary derivative with respect to time. In theories of three-dimensional continua, different notions of time derivation arise depending on whether places are fixed in the reference placement or in the current placement. The material time derivative conventionally arises in this way. When applied to a field in the current placement, it delivers the rate of change in time through two contributions—a *local rate* given by a partial derivative with respect to time and a *convected rate* given by the directional derivative in space along the field of velocity. Since the transplacement (r, F) of a pseudo-rigid body is a function only of t, there is no strict analogue in this theory for the material time derivative. Nevertheless, as we shall see later, its effects are still present here. It is represented again by two contributions. One of these is due to changes in the center of mass and gives rise to the terms on the left-hand sides of $(2.3.13)_1$ and $(2.3.13)_3$. The other arises from changes about the center of mass and leads to the left-hand sides of $(2.3.13)_2$ and $(2.3.13)_4$. These interpretations follow from a precise connection that can be made between the theory of pseudo-rigid bodies and three-dimensional theory. The study of such connections forms the heart of Chapter 3. At this point it is more useful to remain within the foundations of our simple theory and to introduce directly a new notion of time derivation.

The material time derivative arises when we compute time derivatives in the current placement while holding fixed the points of the reference placement. The theory of pseudo-rigid bodies presents a variant of this idea. The Euler functional E and the moment of momentum h are functions on the dual space $\mathscr{L}(\mathscr{V})$, and therefore their values $E(n, k)$ and $h(n)$ change in time, in part owing to the dependence of E and h on time but also owing to any variation in time of n and k. Mirroring the case of the material derivative, we hold fixed here the planes N and K in the reference placement that give rise through transplacement to n and k. This gives rise to a new notion of time derivation peculiar to the theory of pseudo-rigid bodies. Specifically, let A be a (scalar-, vector-, or tensor-valued) function defined on the product space $\mathscr{L}(\mathscr{V}) \times \cdots \times \mathscr{L}(\mathscr{V})$ (p copies). Then the *material-dual derivative* of A, denoted $\mathring{A}$, is that function on the product space that gives the total time derivative of $A(n_1, \ldots, n_p)$ while the preimages of $n_1, \ldots, n_p$ in the reference dual are held fixed. Thus,

$$\mathring{A}(n_1, \ldots, n_p) = \frac{d}{dt}(A(F^{-T}N_1, \ldots, F^{-T}N_p))\Big|_{N_i = F^T n_i}, \qquad (2.3.14)$$

where $N_1, \ldots, N_p$ are all constant.

For the Euler functional and the moment of momentum, this formula becomes

$$\mathring{E}(n, k) = \frac{d}{dt}E(n, k), \qquad \mathring{h}(n) = \frac{d}{dt}h(n), \qquad (2.3.15)$$

where n and k satisfy (2.2.14) with N and K fixed. Since $\mathring{E}$ is a bilinear function and $\mathring{h}$ is linear, we can represent them in terms of tensors $\mathring{E}$ and $\mathring{H}$, respectively, by the relations

$$\mathring{E}(n, k) = n \cdot \mathring{E}k, \qquad \mathring{h}(n) = \mathring{H}n. \qquad (2.3.16)$$

Then (2.3.12) can be reexpressed in the form[8]

$$\mathring{E} = \dot{E} - EL^T - LE, \qquad \mathring{H} = \dot{H} - HL^T. \qquad (2.3.17)$$

Two alternate statements of (2.3.15) provide further insight into the material-dual derivative. For the first, we recall that the current Euler tensor E was introduced so that the values of the Euler functional E would agree, as indicated by (2.2.13), with the values $E_R(N, K)$ of its parent in the reference placement. For the moment of momentum, (2.2.20) expresses a parallel convention. Since N and K are now regarded as constant, we can rewrite (2.3.15) in the form

$$\mathring{E}(n, k) = \dot{E}_R(N, K), \qquad \mathring{h}(n) = \dot{h}_R(N). \qquad (2.3.18)$$

These show that the material-dual derivative arises from carrying the usual ordinary derivative in the reference placement into the current placement of the body, using the same rules by which we carried E_R and h_R to the current placement. In this respect there is a close analogy with the material time derivative of three-dimensional theories.

A second interpretation of (2.3.15) arises when we extend slightly the definition of the material-dual derivative. At present, this derivative applies only to functions defined on a product of dual spaces. Since E and h are not of this kind, our use of the over circle in (2.3.17) is slightly ambiguous. This, however, can easily be resolved by defining $\mathring{E}$ and $\mathring{H}$ directly. Note that the reference placement κ that we have used so far is arbitrary. In particular, we could use as our reference placement the value $\chi(\tau)$ of the motion at some time τ other than the current time t.

---

[8] The material-dual derivative of the Euler tensor was used by COHEN [1981] in his initial study of pseudo-rigid bodies. In particular, $(2.3.17)_1$ is an exact parallel of his (5.21). COHEN also introduced other notions of time derivation and developed a product rule (cf. Eq. (5.26)) and other properties of the material-dual derivative.

With this choice, $(2.2.2)_2$ becomes

$$v = F_{(\tau)}(t) V, \qquad (2.3.19)$$

where $F_{(\tau)}(t)$ is called the *relative deformation*. In terms of the fixed reference placement κ, it is given by

$$F_{(\tau)}(t) = F(t) F(\tau)^{-1}, \qquad (2.3.20)$$

and so (2.2.6) implies that

$$F_{(t)}(t) = I, \qquad \dot{F}_{(t)}(t) = L(t), \qquad (2.3.21)$$

where for the moment an over dot denotes a partial derivative with respect to t. From (2.3.17) and (2.3.21) we find that the material-dual derivative can also be written in the form[9]

$$
\begin{aligned}
\mathring{E}(t) &= \frac{\partial}{\partial t} (F_{(\tau)}(t)^{-1} E(t) F_{(\tau)}(t)^{-\mathrm{T}}) \Big|_{\tau=t}, \\
\mathring{H}(t) &= \frac{\partial}{\partial t} (H(t) F_{(\tau)}(t)^{-\mathrm{T}}) \Big|_{\tau=t}.
\end{aligned}
\qquad (2.3.22)
$$

In order to interpret these results, let $E_{(\tau)}$ and $H_{(\tau)}$ denote the Euler and moment of momentum tensors, respectively, in the reference placement $\chi(\tau)$. The values of these tensors at time t are denoted, respectively, by $E_{(\tau)}(t)$ and $H_{(\tau)}(t)$, just as the values of E and H at time t are $E(t)$ and $H(t)$. If the transformations (2.2.16) and (2.2.21) are specialized to the reference placement $\chi(\tau)$ rather than κ, we find that the expressions in parentheses in (2.3.22) are precisely $E_{(\tau)}(t)$ and $H_{(\tau)}(t)$. Thus, $\mathring{E}(t) = \dot{E}_{(t)}(t)$ and $\mathring{H}(t) = \dot{H}_{(t)}(t)$: *the material-dual derivative is the ordinary time derivative in the reference placement at time τ evaluated at the current time t.* Equivalently, $\mathring{E}$ and $\mathring{H}$ are the rates of change of E and H in time while the current dual space is held instantaneously fixed. As we shall see in Section 2(v), instantaneous rates of this type arise naturally in rigid-body mechanics once body axes and spatial axes are introduced.

Finally, we may reconsider the spatial equations of motion (2.3.13). The material-dual derivative and the conventional ordinary time derivative differ only in connection with quantities defined on products of the dual space $\mathscr{L}(\mathscr{V})$. Since m and p are not of this kind, we may set

$$\mathring{m} = \dot{m}, \qquad \mathring{p} = \dot{p}. \qquad (2.3.23)$$

Thus, (2.3.17) and (2.3.23) show that, in terms of the material-dual derivative, the spatial equations (2.3.13) can be written in the form

---

[9] The results (2.3.17) and (2.3.22) bear some similarity to formulas occurring in the study of materials of the rate type in connection with convected stress rates (cf. TRUESDELL and NOLL [1965, Eqs. (36.17) and (36.20)]). The comparison, however, is not exact, and the interpretations are different.

$$\mathring{m} = 0, \qquad \mathring{E} = 0,$$
$$\mathring{p} = f, \qquad \mathring{H} = M - \Sigma. \tag{2.3.24}$$

From the perspective of the current placement of a body, these provide a closer parallel than (2.3.13) with conventional equations of balance of continuum mechanics.

(iv) Forces and Force-moments

According to the general theory, a pseudo-rigid body is represented by a point in physical space to which is associated some special structure. However, this is only a coarse picture of a real body, and it is useful in the process of modeling external effects to view the body instead as a three-dimensional object of finite size, experiencing concentrated loads and body and surface distributions of force of various kinds. This new picture does not stand in conflict with our developments, since the foundations of the subject say nothing about how we are to obtain the forces and force-moments. For the foundations, external effects are primitive quantities. Modeling them is an enterprise that stands apart from, or in addition to, the theory. Much of what we have said in motivating elements of the theory referred to the points "surrounding" the center of mass, but ultimately such points played no role. Similarly, in modeling f, m_R, and m here we use these surrounding points for motivation, but the quantities we obtain are then assigned directly to the single point that represents the theoretical body.

Consider first the simple case of a body experiencing a number of concentrated forces f_1, f_2, ..., f_p acting at points that in the reference placement are given by places X_1, X_2, ..., X_p. The net force, as suggested by experience from rigid-body mechanics, is simply

$$f = f_1 + f_2 + \cdots + f_p. \tag{2.4.1}$$

Let N be a plane in the reference placement that passes through the center of mass φ_R (cf. Fig. 1). We have interpreted the force-moment $m_R(N)$ as the net moment of all forces with respect to their distance from the plane N. For the force f_i this distance is $(X_i - \varphi_R) \cdot N$, and so multiplying by f_i and forming the net moment over all the forces, we obtain

$$m_R(N) = [(X_1 - \varphi_R) \cdot N]f_1 + [(X_2 - \varphi_R) \cdot N]f_2 + \cdots$$
$$+ [(X_p - \varphi_R) \cdot N]f_p. \tag{2.4.2}$$

In terms of the tensor M_R defined by $(2.3.6)_1$, this can be written equivalently as

$$M_R = f_1 \otimes (X_1 - \varphi_R) + f_2 \otimes (X_2 - \varphi_R) + \cdots + f_p \otimes (X_p - \varphi_R). \tag{2.4.3}$$

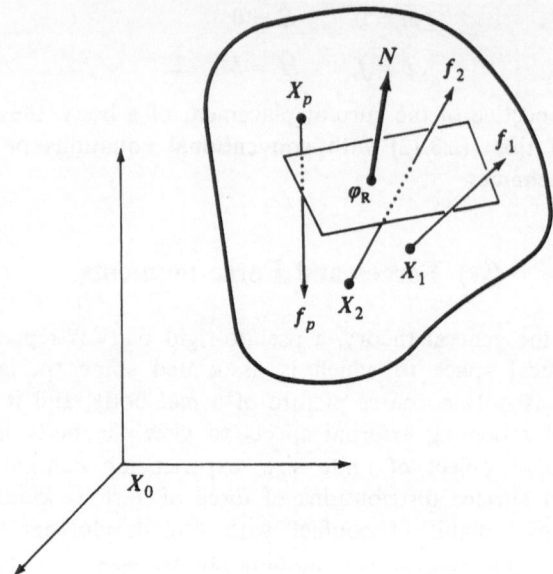

Figure 1

The current force-moment m was defined in terms of m_R by the requirement that the values of the two agree. This is stated explicitly by $(2.3.3)_1$. In order to obtain a more concrete interpretation of m, and its associated tensor M, we recall the transformation $(2.3.8)_1$ between referential and current quantities. With (2.4.3) this gives us

$$M = f_1 \otimes F(X_1 - \varphi_R) + f_2 \otimes F(X_2 - \varphi_R) + \cdots + f_p \otimes F(X_p - \varphi_R), \quad (2.4.4)$$

where we have used the fact that for any vectors a and b, $(a \otimes b)F^T = a \otimes (Fb)$. Next, we note that in motivating various aspects of our theory, we have viewed the vectors v and V in $(2.2.2)_2$ as representations of general position vectors of the matter that surrounds the center of mass, the former in the current placement and the latter in the reference placement. Therefore, we may write $(2.2.2)_2$ equivalently in the form

$$x - \varphi = F(X - \varphi_R), \quad (2.4.5)$$

thus giving us a mapping from reference places to current places. For example, the point of action of the force f_i, viewed from the current placement, is given relative to the mass center by the displacement vector

$$x_i - \varphi = F(X_i - \varphi_R). \quad (2.4.6)$$

Therefore, (2.4.4) can also be written

$$M = f_1 \otimes (x_1 - \varphi) + f_2 \otimes (x_2 - \varphi) + \cdots + f_p \otimes (x_p - \varphi). \quad (2.4.7)$$

This parallels (2.4.3) for the reference placement and provides the following interpretation: for each plane n through φ in the current placement, $m(n)$ represents the net moment of all forces with respect to the distance between n and the current point of action of each force.

It is important to note that the forces $f_1, \ldots, f_p$ are vectors in the current placement of the body, and so we must go to that placement in order to model them. While Fig. 1 shows them in the reference placement, this is purely for convenience since the places at which they act have been specified relative to that placement. Generally the forces change in time, and Fig. 1 only represents their configuration, drawn in the reference placement, at one instant.

In order to illustrate the variety of external effects that can be treated, we consider now some examples and the calculation of f, m_R, and m for each. For simplicity we restrict our attention to a *planar* body, with D_1 and D_2 forming an orthonormal basis for the plane representing the reference placement of the body. As we note in Section 2(v), these vectors can also be interpreted as defining spatial axes for the current placement of the body. The two-dimensional nature of the examples to follow leads to simplicity in the calculations but requires some interpretation. The tensor $D_1 \otimes D_1 + D_2 \otimes D_2$ is the two-dimensional identity, which for the purposes of these examples we denote by I. The tensor F characterizes only deformations in the plane of the body; shears and rotations out of this plane are viewed as absent and there is no deformation along the perpendicular axis. The vector f lies in the plane of the body, and we can interpret it as a three-dimensional force whose component normal to the plane is null; similar interpretations apply to the external force-moment M.

Figure 2 shows a body—circular, of radius a, in the reference placement—with *constant* and *parallel* forces acting at its boundary points $X_1, \ldots, X_4$. These forces might be produced by heavy particles of

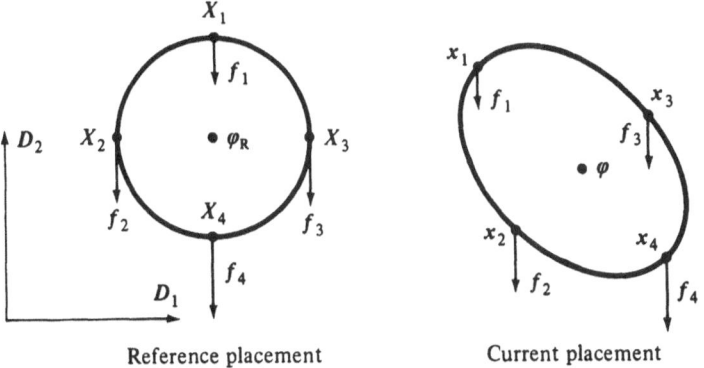

Reference placement   Current placement

Figure 2

various weights attached to the four points. For simplicity we choose

$$f_1 = -wD_2, \qquad f_2 = f_3 = -2wD_2, \qquad f_4 = -3wD_2, \qquad (2.4.8)$$

where w is a positive constant. We have drawn the four forces in the reference placement since they do not change with time. Such forces are called *dead loads*: the force f_i is a function only of the reference point X_i at which it acts. The positions of $X_1, \ldots, X_4$ in Fig. 2, relative to φ_R, are certain multiples of D_1 or D_2. Since the circle has radius a, we have

$$\begin{aligned} X_1 = \varphi_R + aD_2, \qquad X_2 = \varphi_R - aD_1, \\ X_3 = \varphi_R + aD_1, \qquad X_4 = \varphi_R - aD_2. \end{aligned} \qquad (2.4.9)$$

Thus, (2.4.8) and (2.4.9) give us

$$\begin{aligned} f &= -wD_2 - 2wD_2 - 2wD_2 - 3wD_2 = -8wD_2, \\ M_R &= -wD_2 \otimes (aD_2) - 2wD_2 \otimes (-aD_1) \\ &\quad - 2wD_2 \otimes (aD_1) - 3wD_2 \otimes (-aD_2) \\ &= 2waD_2 \otimes D_2. \end{aligned} \qquad (2.4.10)$$

In the current placement, the points of action x_i of the four weights change with time, but their displacement vectors are easily determined through (2.4.6) and (2.4.9) in terms of the deformation F:

$$\begin{aligned} x_1 = \varphi + aFD_2, \qquad x_2 = \varphi - aFD_1, \\ x_3 = \varphi + aFD_1, \qquad x_4 = \varphi - aFD_2. \end{aligned} \qquad (2.4.11)$$

From (2.4.7), (2.4.8), and (2.4.11), we obtain M by a calculation analogous to that giving us M_R:

$$M = 2waD_2 \otimes F(t)D_2. \qquad (2.4.12)$$

Figure 3 shows the same circular body, subject now to forces directed

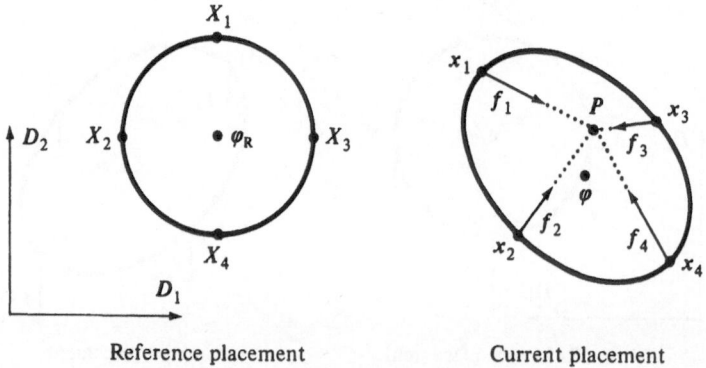

Reference placement   Current placement

Figure 3

toward a fixed point P in the current placement. We assume that the force f_i acts along the line joining P with x_i and is proportional to the distance between the two points:

$$f_i = k(P - x_i), \qquad i = 1, \dots, 4, \tag{2.4.13}$$

where k is a positive constant. Such forces might be produced, for example, by elastic strings stretched tightly between P and each of the places $x_1, \dots, x_4$. In this case, the forces depend only on the current positions of the points of action and thus represent *live loads*. Using the expressions for $x_1, \dots, x_4$ given by (2.4.11), we see from (2.4.13) that

$$f_1 = k(P - \varphi - aFD_2), \qquad f_2 = k(P - \varphi + aFD_1),$$
$$f_3 = k(P - \varphi - aFD_1), \qquad f_4 = k(P - \varphi + aFD_2). \tag{2.4.14}$$

With these, (2.4.1) and (2.4.3) become

$$f = 4k(P - \varphi(t)),$$
$$M_R = -2ka^2\, FD_1 \otimes D_1 - 2ka^2\, FD_2 \otimes D_2 \tag{2.4.15}$$
$$= -2ka^2\, F(t).$$

Here we have used the facts that, in this planar example, $D_1 \otimes D_1 + D_2 \otimes D_2$ is the two-dimensional identity I, and for any vectors a and b, $(Fa) \otimes b = F(a \otimes b)$. By a similar calculation, (2.4.7) reduces to

$$M = -2ka^2\, F(t)F^{\mathrm{T}}(t). \tag{2.4.16}$$

As a final example, consider the situation shown in Fig. 4. In this case, each force f_i is directed from x_i to a fixed place Y_i. The precise

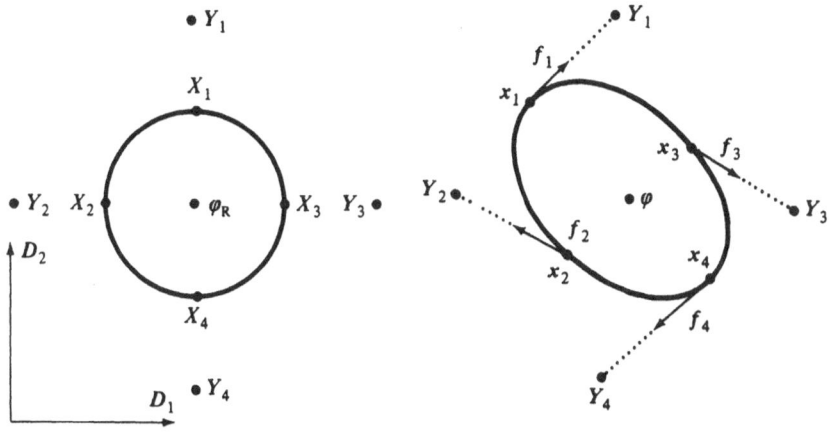

Reference placement   Current placement

Figure 4

locations of these fixed points are given by their representations in the reference placement. Specifically, each Y_i is twice as far from the center of mass φ_R as X_i, and each lies on the line joining X_i and φ_R. Thus,

$$Y_i = \varphi_R + 2(X_i - \varphi_R) = 2X_i - \varphi_R, \qquad i = 1, \ldots, 4, \qquad (2.4.17)$$

or explicitly,

$$Y_1 = \varphi_R + 2aD_2, \qquad Y_2 = \varphi_R - 2aD_1,$$
$$Y_3 = \varphi_R + 2aD_1, \qquad Y_4 = \varphi_R - 2aD_2. \qquad (2.4.18)$$

Once again we assume that f_i is proportional to the difference between Y_i and x_i and is directed toward the fixed point Y_i:

$$f_i = k(Y_i - x_i), \qquad i = 1, \ldots, 4. \qquad (2.4.19)$$

This provides an example of *mixed loads*: by (2.4.17) and (2.4.19) the forces are functions of places in both the reference and current placements. Using (2.4.18) we can write the forces explicitly as

$$f_1 = k(\varphi_R + 2aD_2 - \varphi - aFD_2),$$
$$f_2 = k(\varphi_R - 2aD_1 - \varphi + aFD_1),$$
$$f_3 = k(\varphi_R + 2aD_1 - \varphi - aFD_1),$$
$$f_4 = k(\varphi_R - 2aD_2 - \varphi + aFD_2). \qquad (2.4.20)$$

Now f, M_R, and M can be computed directly:

$$f = 4k(\varphi_R - \varphi(t)),$$
$$M_R = 2ka^2(2I - F(t)), \qquad (2.4.21)$$
$$M = 2ka^2(2I - F(t))F^T(t).$$

In the preceding examples, the external effects vary with time purely through dependence on the points of action $x_1, \ldots, x_4$. Further variety can be obtained by also introducing time explicitly. For example, if the point P in Fig. 3 were not fixed, but instead were made to move about in the current placement according to some function $P = \hat{P}(t)$, then f, M_R, and M would change not only with F but also with t itself.

The expressions (2.4.1), (2.4.3), and (2.4.7) are easily generalized to account for distributions of forces. Let the body occupy in the reference placement a region $\mathcal{B}$, with corresponding referential mass density ρ_R. (For simplicity of notation, in all references to three-dimensional theory we use $\mathcal{B}$ to denote both the body and also the reference region it occupies.) We let b denote the body force per unit mass acting in $\mathcal{B}$, and we let t denote the field of tractions, per unit reference area, acting on $\partial\mathcal{B}$. For the general case of time-dependent mixed loads, we assume b and t are functions of x, X, and t:

$$b = b(x, X, t), \qquad t = t(x, X, t). \qquad (2.4.22)$$

Then the natural generalization of (2.4.1) is

$$f = \int_{\partial \mathcal{B}} t(x, X, t) + \int_{\mathcal{B}} \rho_R b(x, X, t)$$

$$= \int_{\partial \mathcal{B}} t(\varphi + F(X - \varphi_R), X, t) + \int_{\mathcal{B}} \rho_R b(\varphi + F(X - \varphi_R), X, t). \quad (2.4.23)$$

In these expressions, and henceforth in this monograph, integration over $\mathcal{B}$ is with respect to volume and integration over $\partial \mathcal{B}$ is with respect to area, both in the reference placement, and the volume and surface measures are suppressed. It is important to note that x must be replaced by its expression in terms of X, given by (2.4.5), before the integrals are evaluated. This is illustrated explicitly in $(2.4.23)_2$. The corresponding results for M_R and M are

$$M_R = \int_{\partial \mathcal{B}} t(x, X, t) \otimes (X - \varphi_R) + \int_{\mathcal{B}} \rho_R b(x, X, t) \otimes (X - \varphi_R),$$

$$\quad (2.4.24)$$

$$M = \int_{\partial \mathcal{B}} t(x, X, t) \otimes (x - \varphi) + \int_{\mathcal{B}} \rho_R b(x, X, t) \otimes (x - \varphi),$$

where once again x must be replaced by $\varphi + F(X - \varphi_R)$ before the integrals are evaluated.

The explicit examples (2.4.12), (2.4.15), (2.4.16), and (2.4.21) show that f, M_R, and M, while modeling external effects, are often functions of the basic variables φ and F or, by $(2.2.3)_1$, the transplacement (r, F) of the body. This should not be surprising. For the classical example in mass-point mechanics of a mass oscillating on a spring, the external force owing to the spring is precisely a linear function of the current position of the mass. Indeed, this example raises a further possibility not yet considered here. If the effects of air resistance are introduced, the external force on the mass is also a function of velocity. For the case of a pseudo-rigid body then, we might expect that in a very general setting the external effects f, M_R, and M will be functions of r, $\dot{r}$, F, $\dot{F}$, and t. As we shall see in Chapter 3, external effects of this form arise naturally out of a precise interconnection we establish there between the theory of pseudo-rigid bodies and three-dimensional theory.

The force-moments M_R and M serve different purposes in the theory of pseudo-rigid bodies. In the referential equations of motion (2.3.9), the Euler tensor E_R is constant, while in the spatial equations (2.3.13), the Euler tensor E changes in time. As a result, the referential equations are the more useful of the two for solving specific problems. The force-moment M_R is used principally in formulating these equations of motion for specific applications. However, from the perspective of giving physical interpretations of external effects, the appearance of the reference placement is sometimes confusing. These interpretations are clearest when viewed in terms of the current force-moment M. Let us first divide M into its symmetric and skew parts:

$$\text{sym } M = \tfrac{1}{2}(M + M^{\mathrm{T}}), \qquad M_{(ij)} = \tfrac{1}{2}(M_{ij} + M_{ji}),$$
$$\text{sk } M = \tfrac{1}{2}(M - M^{\mathrm{T}}), \qquad M_{[ij]} = \tfrac{1}{2}(M_{ij} - M_{ji}), \tag{2.4.25}$$

where the components are relative to spatial axes defined in the current placement by an orthonormal basis D_1, D_2, D_3. With any skew tensor A, we may associate an axial vector, denoted ax A, such that

$$Av = (\text{ax } A) \times v \tag{2.4.26}$$

for all vectors v. In particular, we have $A(\text{ax } A) = 0$. A simple calculation shows that

$$(a \otimes b - b \otimes a)v = (b \times a) \times v, \tag{2.4.27}$$

and therefore $b \times a = \text{ax}(a \otimes b - b \otimes a)$. Thus, taking the skew part of (2.4.7), we see that

$$\text{ax}(\text{sk } M) = \tfrac{1}{2}\tau, \tag{2.4.28}$$

where

$$\tau = (x_1 - \varphi) \times f_1 + (x_2 - \varphi) \times f_2 + \cdots + (x_p - \varphi) \times f_p. \tag{2.4.29}$$

The vector τ is the classical *resultant torque* about the center of mass.

The tensor sym M is called the *external double force*, and its components represent double forces without moment. For the circular body and system of forces shown in the current placement in Fig. 5, a simple calculation shows that $f = 0$ and

$$M = -2waD_1 \otimes D_1 + 4waD_2 \otimes D_2. \tag{2.4.30}$$

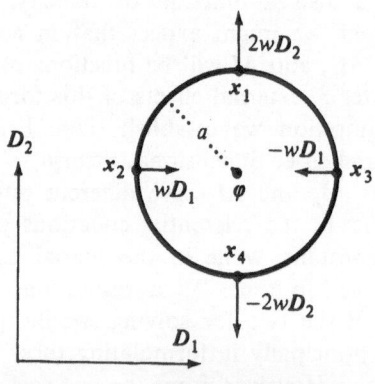

Current placement

Figure 5

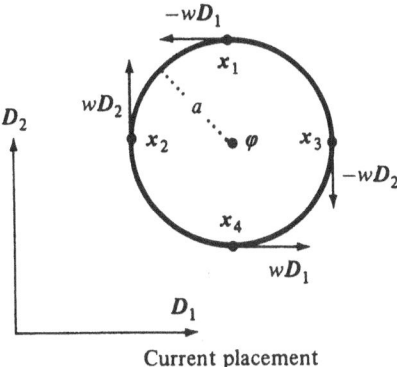

Current placement

Figure 6

Clearly, sk $M = \mathbf{0}$, and it is equally clear that there are no torques present in this example. Also, sym M has only diagonal elements: $M_{(11)}$ represents a compressive force on the body along D_1, $M_{(22)}$ represents a tensile force in the direction D_2. For the same body with the system of forces shown in Fig. 6, f vanishes again and

$$M = -2wa(D_1 \otimes D_2 + D_2 \otimes D_1). \qquad (2.4.31)$$

Since M is symmetric, sk $M = \mathbf{0}$ and once again there is no net torque. In addition, the only nonzero elements of sym M are $M_{(12)} = M_{(21)} = -2wa$, and these clearly produce a shear of the body in the plane of D_1 and D_2.

The components of sk M and sym M thus have direct interpretations in terms of effects producing rotation, extension and compression, and shear. The three explicit examples developed earlier illustrate these interpretations. For the body described by Fig. 2, (2.4.12) shows that generally sk $M \neq \mathbf{0}$, and so in the current placement the body will rotate under the influence of torques. Also, sym M will generally have nontrivial off-diagonal elements, so we may expect shearing in addition to compressive and tensile effects. Note that similar interpretations based on the referential force-moment M_R are completely wrong. By $(2.4.10)_2$, M_R delivers incorrectly only tensile effects in the direction D_2. Shear forces and torques are completely absent.

The body in Fig. 3 presents a different situation. Since the tensor FF^T is symmetric, we see from (2.4.16) that sk $M = \mathbf{0}$. Thus, the body experiences no net torque, though extensions, compressions, and shears are possible in different directions. The absense of torque here is a novel and somewhat surprising result. It indicates that the resultant (2.4.15) of the four forces, which acts at P, must point along the line connecting P to the center of mass φ. Moreover, M_R once again provides the wrong interpretation, incorrectly suggesting the presence of a net torque proportional to sk F.

(v) Rigid-body Mechanics: The Principle of Balance of Angular Momentum

Rigid-body mechanics is founded on two equations of balance. The first is the *balance of linear momentum* $(2.3.13)_3$, in which r denotes the position of the center of mass; the second is the *balance of angular momentum*, which we may write in the classical form

$$\dot{h} = \tau. \tag{2.5.1}$$

The *angular momentum* h of the body is expressed in terms of the *angular velocity vector* w and the inertia tensor J by

$$h = Jw, \tag{2.5.2}$$

and τ is the resultant torque. Equations of balance of mass are often stated as well, though usually in an implicit form: the total mass m and the components of J relative to *body axes* are constant. The former assumption is (2.1.13); the latter can be cast as a differential equation for J, the result being $(2.3.13)_2$ with the Euler tensor E and the deforming tensor L replaced by J and W, respectively.[10] We recall that W is the angular velocity tensor of the body.

It is useful to restate the vector equation (2.5.1) in terms of skew tensors. By (2.4.28) we know that τ is the axial vector $ax(M - M^T)$, where M is the current force-moment tensor. Similarly, we may show that h is the axial vector $ax(H - H^T)$, where

$$H = WE. \tag{2.5.3}$$

The proof appeals to the identity[11]

$$(\operatorname{tr} E)[v, u, z] = [Ev, u, z] + [v, Eu, z] + [v, u, Ez], \tag{2.5.4}$$

which is valid for any symmetric tensor E and any vectors v, u, and z. The triple scalar product $[v, u, z]$ is defined by

$$[v, u, z] = v \cdot u \times z. \tag{2.5.5}$$

Since the angular velocity w is the axial vector $ax\ W$, we conclude from (2.1.21), (2.4.26), (2.5.3), and the above identity that, for any vectors u and v,

---

[10] See, for example, JAUNZEMIS [1967, Eq. (9.51)]. Within the theory of pseudo-rigid bodies itself, we can generate from (2.1.21) and $(2.3.13)_2$ a differential equation for J that parallels $(2.3.13)_2$ for E. Unfortunately, it is sufficiently complicated to serve little value, simplifying to a tractable form only when L is skew.

[11] JAUNZEMIS [1967, Eq. $(4.7)_1$].

$$u \cdot (H - H^T)v = u \cdot (WE + EW)v$$
$$= -Wu \cdot Ev + Eu \cdot Wv$$
$$= -(w \times u) \cdot Ev + Eu \cdot (w \times v)$$
$$= [Ev, u, w] + [v, Eu, w]$$
$$= (\operatorname{tr} E)[v, u, w] - [v, u, Ew]$$
$$= [u, (\operatorname{tr} E)w - Ew, v]$$
$$= u \cdot (Jw \times v). \tag{2.5.6}$$

Since u is arbitrary, we see from (2.5.2) that

$$(H - H^T)v = h \times v, \tag{2.5.7}$$

and this completes the proof. We conclude that the balance of angular momentum for a rigid body can be expressed in the equivalent tensor form

$$(\operatorname{sk} H)^{\cdot} = \operatorname{sk} M. \tag{2.5.8}$$

The theory of pseudo-rigid bodies has been designed to provide a generalization of rigid-body mechanics. At the same time it should also give a coarse description of the motions of three-dimensional bodies. Whether it succeeds in either of these respects is a matter for study rather than conjecture. Studies of this type—whether one theory of material behavior can be considered a sub- or supertheory of another— are rare in continuum mechanics. We call this the question of *consistency*, and in this monograph it arises explicitly in two different contexts. In the first, which is developed in Chapter 3, we begin with the theory of three-dimensional nonlinear elasticity, and we study in depth whether the theory of pseudo-rigid bodies represents for it a consistent *subtheory*. While a precise definition of that term must wait until Chapter 3, a subtheory can broadly be viewed as a subset of solutions for a given theory that forms, within itself, a separate theory. The second study of the question of consistency is somewhat simpler and is related directly to our present concerns. Namely, *does the classical mechanics of rigid bodies form a consistent subtheory of the mechanics of pseudo-rigid bodies?* It is natural[12] to look to the subset of rigid motions of a pseudo-rigid body as the set of solutions representing the subtheory. In this case, the question can be rephrased as follows: *Are the rigid motions of a pseudo-rigid body completely determined by the classical balance principles*

---

[12] It might be argued that rigid-body mechanics, being a theory of a constrained continuum, would arise more naturally out of the theory of pseudo-rigid bodies as the limit of a sequence of pseudo-rigid continua, each more resistant to deformation than its predecessor. Chapter 6 explores this approach in detail in terms of the concept of an *almost-rigid body*.

$(2.3.13)_3$ and $(2.5.1)$, *which govern the motions of rigid bodies*? If the answer is yes, then we may view rigid-body mechanics as a *consistent subtheory* of the mechanics of pseudo-rigid bodies.

These two issues of consistency, one relating to the study of pseudo-rigid bodies through nonlinear elasticity and the other relating to the study of rigid bodies through pseudo-rigid body mechanics, must be approached in different ways. In the former case, nonlinear elasticity is viewed as a fixed, well-recognized theory, whose predictions about material behavior are much finer in their detail than those of the theory we are developing here. If the mechanics of pseudo-rigid bodies is a subtheory of elasticity theory, we must demonstrate this through analysis and proof. Chapter 3 is concerned completely with this problem. The second case, however, presents a different question. The mechanics of pseudo-rigid bodies now represents the fine theory, at this point still in the process of formulation. It is quite natural to view consistency with rigid-body mechanics as an *a priori requirement* and therefore to impose it directly as an additional axiom. Thus, we ensure consistency in this case by *arrangement* rather than establishing it by analysis.

We wish to require then that the balance of angular momentum (2.5.8) is a consequence of our theory in any rigid motion of a pseudo-rigid body. We impose in fact more than this. It is quite natural to include in continuum theories a balance of angular momentum. Since the current moment of momentum tensor H given by $(2.2.28)_2$ reduces exactly to (2.5.3) in a rigid motion, we define the *angular momentum A* of a pseudo-rigid body to be

$$A = \text{sk } H = \text{sk}(LE). \qquad (2.5.9)$$

We postulate then that *in any pseudo-rigid motion the equation of balance of angular momentum applies*:

$$\dot{A} = \text{sk } M. \qquad (2.5.10)$$

In particular, (2.4.28) shows that the angular momentum A is constant in any torque-free motion of a pseudo-rigid body.

Since LEL^{T} is symmetric, we note from (2.5.9) and $(2.3.17)_2$ that the ordinary time derivative and the material-dual derivative agree for the angular momentum:

$$\overset{\circ}{A} = \dot{A}. \qquad (2.5.11)$$

This provides one explanation for why the material-dual derivative does not arise in the mechanics of rigid bodies. If we compare (2.5.10) with the skew part of the moment of momentum balance $(2.3.24)_4$, we see from (2.5.11) that the principle of balance of angular momentum is equivalent to

$$\Sigma = \Sigma^{\mathsf{T}}: \qquad (2.5.12)$$

the current internal force-moment Σ is symmetric. In view of the interpretations given in Section 2(iv) to the components of the external force-
moment M, we henceforth call Σ the *internal double force.* We see here
a direct analogy with three-dimensional theory: the balance of angular
momentum is conventionally phrased in terms of the symmetry of the
Cauchy stress. Indeed, as we shall see in Chapter 3, for consistency
between nonlinear elasticity and the mechanics of pseudo-rigid bodies,
Σ must be proportional to *the average Cauchy stress* over the current
placement of the body. The analogy can be carried further by noting
that (2.5.12), when rephrased through (2.3.8)$_2$ in terms of the reference
placement, becomes

$$\Sigma_R F^T = F \Sigma_R^T, \tag{2.5.13}$$

a relation paralleling that satisfied by the first Piola–Kirchhoff stress
tensor (cf. TRUESDELL and NOLL [1965, Eq. (43A.8)]).

With the postulate (2.5.10), rigid-body mechanics becomes a subtheory
of the mechanics of pseudo-rigid bodies. Structurally, however, the two
theories still differ in several respects. The most important of these relates
to the role of the reference placement. The use of a reference placement in classical physics in neither emphasized nor commonplace, but in
the theory proposed here it plays a distinguished role. The difference,
however, is only superficial. Through the introduction of spatial and
body axes, and representations of variables relative to each, rigid-body
mechanics achieves much that the use of a reference placement provides.
While an in-depth study of body axes, suitably generalized to the case
of a pseudo-rigid body, is not essential to our theory, it is nevertheless useful in drawing out parallels with some results peculiar to the
mechanics of rigid bodies.

Since the basic kinematics of the director space $\mathscr{D}$ is restricted to
invertible linear transformations, we can represent this kinematics equivalently in terms of a basis for $\mathscr{D}$. The reference placement κ transforms
this basis into a corresponding basis D_i, $i = 1, 2, 3$, for $\mathscr{V}$. Under the
action of any transplacement, this second basis is mapped into a certain
time-dependent basis in the current placement, denoted $d_i(t)$, $i = 1, 2, 3$,
and a motion of the pseudo-rigid body can then be represented by

$$q = r(t), \qquad d_i = F(t)D_i, \qquad i = 1, 2, 3. \tag{2.5.14}$$

Let D^i and d^i, $i = 1, 2, 3$, be bases for the dual space $\mathscr{L}(\mathscr{V})$ in the
reference and current placements, respectively, and choose these to be
dual to the bases in $\mathscr{V}$. Thus,[13]

$$D^i \cdot D_j = \delta_j^i, \qquad d^i \cdot d_j = \delta_j^i. \tag{2.5.15}$$

---

[13] δ_j^i denotes the Kronecker delta, and the summation convention applies to all repeated
indices.

According to (2.2.14), specialized to elements of the bases for $\mathscr{L}(\mathscr{V})$, $\boldsymbol{D}^i$ and $\boldsymbol{d}^i$ are related by

$$\boldsymbol{d}^i(t) = F(t)^{-\mathrm{T}} \boldsymbol{D}^i, \qquad i = 1, 2, 3. \tag{2.5.16}$$

The vectors $\boldsymbol{d}_i$ are called the *directors* of the pseudo-rigid body, and in terms of them the deformation becomes

$$F(t) = \boldsymbol{d}_i(t) \otimes \boldsymbol{D}^i. \tag{2.5.17}$$

In this format, a pseudo-rigid body is viewed as a zero-dimensional body with three directors, an interpretation initially proposed by MUNCASTER [1984b].

Spatial axes are a set of axes fixed in space in the current placement of the body. If we formally identify the vector space $\mathscr{V}$ in the reference placement with this same space in the current placement, we may view the basis vectors $\boldsymbol{D}_i$ and their duals $\boldsymbol{D}^i$ as defining a set of axes fixed in the current placement and its dual. We call these the *spatial axes* of the pseudo-rigid body. In contrast, the directors $\boldsymbol{d}_i$ and $\boldsymbol{d}^i$ given by $(2.5.14)_2$ and (2.5.16) move naturally with the body, and we may view these as defining a set of *body axes*. From the perspective of an observer fixed in space, the vectors $\boldsymbol{D}_i$ and their duals $\boldsymbol{D}^i$ are constant, and in this case $(2.5.14)_2$, (2.5.16), and (2.2.6) imply that

$$\begin{aligned}
\dot{\boldsymbol{d}}_i &= \dot{F}F^{-1}\boldsymbol{d}_i = L\boldsymbol{d}_i, \\
\dot{\boldsymbol{d}}^i &= -(F^{-1}\dot{F}F^{-1})^{\mathrm{T}}F^{\mathrm{T}}\boldsymbol{d}^i = -L^{\mathrm{T}}\boldsymbol{d}^i.
\end{aligned} \tag{2.5.18}$$

Physically, these indicate that the deforming tensor L measures the rate of change of the body axes relative to the fixed spatial axes.

Consider now the representations of our basic variables relative to body axes. Since E, H, M, and Σ are transformations from $\mathscr{L}(\mathscr{V})$ to $\mathscr{V}$, we express them in terms of the tensor-basis elements $\boldsymbol{d}_i \otimes \boldsymbol{d}_j$. The deforming tensor L maps $\mathscr{V}$ to $\mathscr{V}$, so for it we use $\boldsymbol{d}_i \otimes \boldsymbol{d}^j$. Thus, we may write

$$\begin{aligned}
E &= E_{\mathrm{R}}^{ij}\boldsymbol{d}_i \otimes \boldsymbol{d}_j, \qquad H = H^{ij}\boldsymbol{d}_i \otimes \boldsymbol{d}_j, \\
M &= M^{ij}\boldsymbol{d}_i \otimes \boldsymbol{d}_j, \qquad \Sigma = \Sigma^{ij}\boldsymbol{d}_i \otimes \boldsymbol{d}_j, \\
L &= L_j^i\boldsymbol{d}_i \otimes \boldsymbol{d}^j.
\end{aligned} \tag{2.5.19}$$

The subscript R on the components of E is more than coincidental. Using (2.5.16) and (2.2.16), we find that

$$\begin{aligned}
E_{\mathrm{R}}^{ij} &= \boldsymbol{d}^i \cdot E\boldsymbol{d}^j \\
&= F^{-\mathrm{T}}\boldsymbol{D}^i \cdot EF^{-\mathrm{T}}\boldsymbol{D}^j \\
&= \boldsymbol{D}^i \cdot F^{-1}EF^{-\mathrm{T}}\boldsymbol{D}^j \\
&= \boldsymbol{D}^i \cdot E_{\mathrm{R}}\boldsymbol{D}^j.
\end{aligned} \tag{2.5.20}$$

Thus, *the components of the current Euler tensor E relative to body axes are constants*; they equal the components of the referential Euler tensor E_R relative to the basis fixed in the reference placement. This fact is the precise reason that body axes are used in the mechanics of rigid bodies, and it highlights the close parallel between them and the use of a reference placement.

At the same time, body axes introduce certain complications. We may interpret the components appearing in (2.5.19) as those of an observer moving with the body. "Observer" in this context goes beyond the usual meaning of that term from classical physics. Since the body is not only translating and rotating but also deforming relative to the fixed spatial axes, we must view the observer as experiencing similar effects. Nevertheless, from the perspective of this observer, the basis elements d_i and d^i are not changing in time. Since the over dot conventionally refers to a time rate relative to axes fixed in space, rates as calculated by the moving observer will generally be different. It is useful to introduce these explicitly. Let $P(t)$ denote a vector-valued function of time. We define its *codeforming derivative* $\overset{\circ}{P}$ to be

$$\overset{\circ}{P}(t) = \frac{\partial}{\partial t}(F_{(\tau)}(t)^{-1}P(t))\Big|_{\tau=t}. \qquad (2.5.21)$$

In view of (2.3.20) and (2.3.21), we obtain the explicit formula

$$\overset{\circ}{P} = \dot{P} - LP. \qquad (2.5.22)$$

Writing this as

$$\dot{P} = \overset{\circ}{P} + LP \qquad (2.5.23)$$

and noting that in a rigid motion $LP = WP = w \times P$, we recover a formula from rigid-body mechanics[14] that shows that $\overset{\circ}{P}$ is the time rate of change of P relative to an observer moving with the body. This interpretation follows directly from (2.5.21). The vector $P(t)$ is momentarily pulled back to the reference placement $\chi(\tau)$, differentiated with respect to t holding this placement fixed, and the result is instantaneously carried forward again to the current placement $\chi(t)$. Alternately, if

$$P = P^i d_i \qquad (2.5.24)$$

is the component form of P relative to body axes, then (2.5.22) and (2.5.18)$_1$ show that

$$\overset{\circ}{P} = \dot{P}^i d_i, \qquad (2.5.25)$$

and so $\overset{\circ}{P}$ is just the time derivative of P while holding the d_i fixed. By (2.5.18)$_1$, the extra term LP in (2.5.23) is the contribution to the absolute time rate $\dot{P}$ owing to the motion of the body axes relative to the spatial axes.

---

[14] SYNGE [1959, Eq. (12.306)], but with $\overset{\circ}{P}$ written as $\delta P/\delta t$.

Codeforming derivatives of H and L can be defined in a similar way. For these we must pull back both the domain and range of each mapping to $\chi(\tau)$, differentiate with respect to t, and then carry the results forward again to $\chi(t)$. For H this gives us

$$\mathring{H} = \frac{\partial}{\partial t}(F_{(\tau)}(t)^{-1} H(t) F_{(\tau)}(t)^{-T})\bigg|_{\tau=t}, \qquad (2.5.26)$$

while for L we have

$$\mathring{L} = \frac{\partial}{\partial t}(F_{(\tau)}(t)^{-1} L(t) F_{(\tau)}(t))\bigg|_{\tau=t}. \qquad (2.5.27)$$

Using (2.3.21) we can compute these in explicit tensor form:

$$\mathring{H} = \dot{H} - LH - HL^T, \qquad \mathring{L} = \dot{L} - LL + LL = \dot{L}. \qquad (2.5.28)$$

In terms of the component representations (2.5.19), they become

$$\mathring{H} = \mathring{H}^{ij} d_i \otimes d_j, \qquad \mathring{L} = \mathring{L}^i_j d_i \otimes d^j. \qquad (2.5.29)$$

By $(2.3.17)_2$, $(2.5.28)_1$, and $(2.2.28)_2$, we see that the material-dual derivative of H has the form

$$\mathring{H} = \dot{H} + LH$$
$$= (LE)^{\boldsymbol{\cdot}} + L^2 E$$
$$= (\mathring{L} + L^2)E, \qquad (2.5.30)$$

where in the last step we have used the fact that E, relative to body axes, is constant.[15] Thus, the equation of balance $(2.3.24)_4$ becomes

$$(\mathring{L} + L^2)E = M - \Sigma. \qquad (2.5.31)$$

By (2.5.29), the component form of this equation is the system of differential equations

$$(\mathring{L}^i_p + L^i_q L^q_p)E_R^{pj} = M^{ij} - \Sigma^{ij}. \qquad (2.5.32)$$

At this point a more explicit comparison with rigid-body mechanics is useful. Set $L = W$ and consider the skew part of (2.5.31). By the symmetry of the internal double force Σ, we obtain

$$(\mathring{W}E + E\mathring{W}) + (W^2 E - EW^2) = M - M^T. \qquad (2.5.33)$$

The axial vector corresponding to the right-hand side is precisely the net torque τ. The axial vector corresponding to $\mathring{W}$ is the time derivative, relative to body axes, of the angular velocity vector w. If we denote this time derivative by $\mathring{w}$, the calculation (2.5.6), applied now to $\mathring{H}$ rather than H, shows that the axial vector corresponding to the first term on

---

[15] Moreover, it is easy to show that $\mathring{E} = \dot{E}$, provided we define $\mathring{E}$ by a formula parallel to (2.5.26).

the left-hand side of (2.5.33) is $J\overset{\circ}{w}$. Finally, using the identity

$$u \times (v \times z) = (u \cdot z)v - (u \cdot v)z, \qquad (2.5.34)$$

we see from (2.1.21) that for any vector v,

$$
\begin{aligned}
(W^2 E - EW^2)v &= (JW^2 - W^2 J)v \\
&= J[w \times (w \times v)] - w \times (w \times Jv) \\
&= J[(w \cdot v)w - (w \cdot w)v] - [(w \cdot Jv)w - (w \cdot w)Jv] \\
&= (w \cdot v)Jw - (Jw \cdot v)w \\
&= v \times (Jw \times w) \\
&= (w \times Jw) \times v. \qquad (2.5.35)
\end{aligned}
$$

Therefore, since (2.5.22) easily shows $\overset{\circ}{w} = \dot{w}$, the axial vector equation corresponding to (2.5.33) is

$$J\overset{\circ}{w} + w \times Jw = \tau \quad \text{or} \quad J\dot{w} + w \times Jw = \tau. \qquad (2.5.36)$$

The component form of this equation relative to body axes gives precisely EULER'S *equations for a rigid body*. This shows that (2.5.31) and (2.5.32) provide direct generalizations of EULER'S classical equations to the theory of pseudo-rigid bodies, the first in tensor form and the second in terms of components.

The foregoing results have been obtained through undue cost and elaboration. In modern continuum mechanics the use of body axes is neither essential nor encouraged, and the discussion here is an excellent illustration. Indeed, differentiation of (2.2.6) shows that $\dot{F} = (\dot{L} + L^2)F$, and so the referential equation (2.3.9)$_4$ becomes

$$(\dot{L} + L^2)FE_R = M_R - \Sigma_R. \qquad (2.5.37)$$

The component form of this equation relative to the basis $d_i \otimes D_j$ is precisely the system (2.5.32) once we observe, from (2.5.17), that relative to the basis $d_i \otimes D^j$ the representation of the deformation F is the identity matrix. Alternately, if we multiply (2.5.37) on the right by F^T, we see by (2.2.16), (2.3.8)$_{1,2}$, and (2.5.28)$_2$ that the result is exactly the coordinate-free equation (2.5.31).

(vi) Changes of Frame

Previously we introduced the extended event world $\mathscr{W}_e$ and defined a frame of reference on it by the decomposition (2.1.5) into spatial and temporal components. The choice of the frame of reference on $\mathscr{W}_e$, as reflected in this decomposition, is not unique. It $\mathscr{E}^*$ and $\mathscr{I}^*$ are Euclidean spaces isomorphic to, but possibly distinct from, $\mathscr{E}$ and $\mathscr{I}$, respectively, then the isomorphism

$$\mathcal{W}_e \cong \mathcal{E}^* \times \mathcal{V}^* \times \mathcal{I}^* \tag{2.6.1}$$

would serve equally well as a frame of reference. Naturally, $\mathcal{V}^*$ represents the translation space of $\mathcal{E}^*$. In physics we refer to a new choice of the decomposition of $\mathcal{W}_e$ as a *change of observer*. Any event $w \in \mathcal{W}_e$ now has the two coordinate representations (x, v, i) and (x^*, v^*, i^*), and a change of observer can be specified concretely as a point mapping from one of these to the other. However, only selected mappings, as characterized by a pair of transformations, are interesting physically. The first member of this pair is a point transformation from $\mathcal{I}$ to $\mathcal{I}^*$, which we may interpret as a reparameterization of time. The second element is a transformation from $\mathcal{I} \times \mathcal{E}$ to $\mathcal{E}^*$, which we may view as a time-dependent point transformation between the physical spaces $\mathcal{E}$ and $\mathcal{E}^*$; it models movement in space of one observer relative to the other. Of course, there is an associated transformation from $\mathcal{I} \times \mathcal{V}$ to $\mathcal{V}^*$ induced on the translation spaces. When the transformation from $\mathcal{E}$ to $\mathcal{E}^*$ for each time is an isometry, and when the transformation from $\mathcal{I}$ to $\mathcal{I}^*$ is an orientation-preserving isometry, we call the change of observer a *change of frame*.

A theorem in geometry[16] shows that in a change of frame the mapping from $\mathcal{E}$ to $\mathcal{E}^*$ for each time t corresponds to a *translation* combined with an *orthogonal transformation*. Moreover, the transformation from $\mathcal{V}$ to $\mathcal{V}^*$ for each time is also *orthogonal*, and the mapping from $\mathcal{I}$ to $\mathcal{I}^*$ is a *translation*. If $z(t)$ is an arbitrary time-dependent place in $\mathcal{E}^*$, and if $Q(t)$ is an arbitrary time-dependent orthogonal transformation from $\mathcal{V}$ to $\mathcal{V}^*$, then we can represent a change of frame by the equations

$$
\begin{aligned}
x^* &= z(t) + Q(t)(x - x_0), \\
v^* &= Q(t)v, \\
t^* &= t_0 + t.
\end{aligned}
\tag{2.6.2}
$$

The place x_0 is the fixed origin in $\mathcal{E}$ introduced earlier, and x_0^* will denote a fixed origin in $\mathcal{E}^*$. t_0 is an arbitrary constant, and t and t^* are the coordinates of i and i^* relative to fixed time origins i_0 and i_0^*, respectively.

The point transformation (2.6.2) leads naturally to a corresponding transformation of pseudo-rigid motions. By combining (2.1.11) with (2.6.2), we find that the new center of mass φ^* and the new transformation $\Phi^* \in \mathcal{I}nv(\mathcal{D}, \mathcal{V}^*)$ are given by

$$
\begin{aligned}
\varphi^*(P, t^*) &= z(t) + Q(t)(\varphi(P, t) - x_0), \\
v^* &= \Phi^*(P, t^*)V \\
&= Q(t)\Phi(P, t)V.
\end{aligned}
\tag{2.6.3}
$$

---

[16] NOLL [1964].

These formulas are simplest to interpret if, instead of viewing (2.6.2) and (2.6.3) as transformations between different spaces, we regard (2.6.3) as a mapping between two different motions in the same space.[17] Then we obtain the simpler representation

$$\varphi^*(P, t^*) = \varphi(P, t) + k(t),$$

$$\Phi^*(P, t^*) = Q(t)\Phi(P, t),$$
(2.6.4)

where

$$k(t) = z(t) - \varphi(P, t) + Q(t)(\varphi(P, t) - x_0).$$
(2.6.5)

k can be viewed as an arbitrary time-dependent translation of the center of mass and $Q(t)$ as an arbitrary time-dependent orthogonal transformation of $\mathcal{V}$. Equivalently, by (2.2.3) we can write this as

$$r^*(t^*) = r(t) + k(t) + x_0 - x_0^*, \qquad F^*(t^*) = Q(t)F(t).$$
(2.6.6)

In this form we see that the transformation splits in a natural way. A pseudo-rigid body is described in terms of the location of its mass center r and also in terms of neighboring points through the deformation F. A change of frame, by (2.6.6), translates the center of mass and applies an orthogonal transformation to points near it. Even though the orthogonal transformation Q appears in (2.6.5), since we are applying the general point transformation (2.6.2)$_1$ to a single place, namely, the center of mass, the overall effect is simply the translation defined by $k + x_0 - x_0^*$.

If a pseudo-rigid body is undeformed relative to the unstarred frame, then $F = I$ and r is a fixed constant r_0, and the transformed motion becomes

$$r^*(t) = r_0 + k(t) + x_0 - x_0^*, \qquad F^*(t) = Q(t).$$
(2.6.7)

Here and henceforth we set $t_0 = 0$, so that t and t^* agree. We call (2.6.7) a *rigid-body motion* of $\mathcal{B}_e$. If k and Q are independent of time, it is called a *rigid-body displacement*. Physical considerations might lead one to require that the orthogonal transformation Q be a rotation, that is, $Q \in \mathcal{O}^+(\mathcal{V})$, but generally we shall not impose this restriction. We observe now that the general motion defined by the right-hand sides of (2.6.6) is a composition of two; specifically, the transformation (2.6.6) can be viewed as the process of imparting a *superposed rigid-body motion* (or displacement, if k and Q are constant) to the given motion (r, F).

In mechanics we deal with objects, such as scalars, vectors, tensors, and their generalizations, which obey special rules of transformation under a change of frame. These rules are founded on the requirement that a quantity have a meaning that is independent of the choice of observer, thereby making it intrinsic to the event world $\mathcal{W}_e$ itself. This requirement is called the *principle of frame indifference*. A scalar s is

---

[17] This is the *passive* to *active* transition used by SALETAN and CROMER [1971, p. 68].

frame indifferent if its value is the same for all observers. Thus,

$$s^* = s. \tag{2.6.8}$$

A vector v is frame indifferent if it transforms as does an element of the translation space $\mathscr{V}$, that is, according to $(2.6.2)_2$. The rules of transformation for most other quantities follow from, or are variants of, these two. For example, a tensor $T \in \mathscr{V} \otimes \mathscr{V}$ can be viewed as a linear mapping of vectors onto vectors. In order for T to be frame indifferent, we define T^* as that linear mapping such that, for all frame-indifferent vectors u and v, $v^* = T^* u^*$ whenever $v = Tu$. This implies that

$$T^* = QTQ^{\mathrm{T}}. \tag{2.6.9}$$

A second example is provided by the rule of transformation for elements of a dual space. A linear functional $n \in \mathscr{L}(\mathscr{V})$ is important only in respect to its action on vectors. We say that n is frame indifferent if its transform n^* is such that $n^* \cdot v^* = n \cdot v$ for all frame-indifferent vectors v. That is, $n \cdot v$ is a frame-indifferent scalar. Since $n \cdot v = n \cdot Q^{\mathrm{T}} Q v = Qn \cdot Qv = Qn \cdot v^*$, we find that

$$n^* = Qn. \tag{2.6.10}$$

The requirement of frame indifference can be extended to place-valued, scalar-valued, vector-valued, etc., functions of places, scalars, vectors, etc. The guiding rule for functions says that the *value* of a function must transform as a frame-indifferent quantity whenever the *arguments* of the function transform as frame-indifferent quantities. As an example, consider the current Euler functional E. Since it is a scalar-valued function on $\mathscr{L}(\mathscr{V}) \times \mathscr{L}(\mathscr{V})$, we define E^* by the requirement that $E^*(n^*, k^*) = E(n, k)$ for all frame-indifferent planes n and k. By (2.6.10) we find that the *function* E^* is defined by

$$E^*(n, k) = E(Q^{\mathrm{T}} n, Q^{\mathrm{T}} k). \tag{2.6.11}$$

Similarly, the current moment of momentum h and the current force-moment m are vector-valued functions on $\mathscr{L}(\mathscr{V})$, and so we require that $h^*(n^*) = Qh(n)$ and $m^*(n^*) = Qm(n)$ for all frame-indifferent planes n. These imply that the *functions* h^* and m^* are

$$h^*(n) = Qh(Q^{\mathrm{T}} n), \qquad m^*(n) = Qm(Q^{\mathrm{T}} n). \tag{2.6.12}$$

Previously we expressed E, h, and m in terms of tensors through (2.1.20), (2.2.19), and $(2.3.7)_1$. In order that these representations apply to all observers, we find from (2.6.11) and (2.6.12) that the tensors corresponding to E^*, h^*, and m^* are given by

$$E^* = QEQ^{\mathrm{T}}, \qquad H^* = QHQ^{\mathrm{T}}, \qquad M^* = QMQ^{\mathrm{T}}. \tag{2.6.13}$$

Thus, E, H, and M are frame-indifferent tensors.

It is important to note that we are considering changes of frame only in regard to the current placement of the body. The reference placement has been chosen and fixed, independent of the transformations (2.6.2) that apply to current places and vectors. Thus, reference vectors V and reference planes N remain the same under a change of frame. As a result, functions defined either on the reference space or its dual have transforms slightly different from those given above. For example, the deformation F is a mapping from *referential* to *current* vectors. Frame indifference then suggests that we define F^* so that $v^* = F^*V$ if $v = FV$ whenever v is a frame-indifferent vector. By $(2.6.2)_2$ we see that

$$F^* = QF, \qquad (2.6.14)$$

and therefore F transforms as a frame-indifferent vector. The Euler functional E_R, moment of momentum h_R, and force-moment m_R are all functions defined only on the set of reference planes. Since this set is fixed, the first is simply a frame-indifferent scalar-valued function, while the second two are frame-indifferent vector-valued functions:

$$E_R^*(N, K) = E_R(N, K), \qquad h_R^*(N) = Qh_R(N), \qquad m_R^*(N) = Qm_R(N). \quad (2.6.15)$$

Similarly, the tensors corresponding to these transform under a change of frame according to

$$E_R^* = E_R, \qquad H_R^* = QH_R, \qquad M_R^* = QM_R. \qquad (2.6.16)$$

In Chapter 3, where we examine comparisons with the theory of nonlinear elasticity, frame indifference arises in connection with functions defined on the set of places and times. For example, the field of body force b per unit mass is viewed as a vector-valued function $b(t, x)$ of times t and places $x \in \mathscr{E}$. Under a change of frame, an observer sees a new *body-force function* b^* defined on $\mathscr{I}^* \times \mathscr{E}^*$. Its value $b^*(t^*, x^*)$ must be the vector transform of $b(t, x)$. Thus,

$$b^*(t^*, x^*) = Qb(t, x), \qquad (2.6.17)$$

where x^* and x are related by $(2.6.2)_1$, and $t^* = t$ according to our previous convention. This relation explicitly defines the function b^*. Indeed, solving $(2.6.2)_1$ for x in terms for x^*, we obtain

$$x = Q^T(t)(x^* - z(t)) + x_0, \qquad (2.6.18)$$

and substituting this into (2.6.17) and replacing x^* with x, we arrive at the formula

$$b^*(t, x) = Q(t)b(t, Q^T(t)(x - z(t)) + x_0). \qquad (2.6.19)$$

In particular, if b is independent of t and the change of frame is a rigid-body displacement, then b^* is also independent of t.

(vii) Constitutive Relations

The constitutive relations (2.3.1) and (2.3.4), or their equivalent tensor forms $(2.3.6)_3$ and $(2.3.7)_3$, define the type of material we choose to consider. By our choice of any one of the response functions $\mathfrak{s}_R$, $\mathfrak{S}_R$, $\mathfrak{s}$, or $\mathfrak{S}$ we may focus on any of a wide variety of material response, from rate-dependent materials to anisotropic elastic solids, and even elastic fluids. Our objective here, however, is to obtain a theory with strong practical implications, and so it is reasonable to focus only on the simplest and most well-developed materials—isotropic elastic and hyperelastic solids. We leave the consideration of more exotic materials to future studies.

An *elastic* pseudo-rigid body is defined by either of the following two equivalent constitutive relations:

$$\sigma_R = \mathfrak{s}_R(F, N) = \mathfrak{S}_R(F)N \quad \text{or} \quad \sigma = \mathfrak{s}(F, n) = \mathfrak{S}(F)n. \qquad (2.7.1)$$

They state that the internal force-moment is determined by the current value of the deformation F rather than by its complete history. In tensor form,

$$\Sigma_R = \mathfrak{S}_R(F) \quad \text{or} \quad \Sigma = \mathfrak{S}(F), \qquad (2.7.2)$$

where we recall the transformations $(2.3.8)_{2,3}$ that carry the first into the second. A *hyperelastic* pseudo-rigid body is defined by the constitutive relation

$$\Sigma_R = \frac{\partial \sigma}{\partial F}(F) \quad \text{or} \quad \Sigma = \frac{\partial \sigma}{\partial F}(F)F^T, \qquad (2.7.3)$$

where $\sigma: \mathcal{Gl}(\mathcal{V}) \to \mathcal{R}$ is called the *stored-energy function*. Here and henceforth $\mathcal{R}$ denotes the set of real numbers. The class of hyperelastic materials comprises a subset of the class of elastic materials.

We noted previously, in connection with (2.3.1), that response functions generally depend on the choice of the reference placement κ. This is often reflected through special properties of $\mathfrak{S}_R$ and σ that hold for one choice of κ but not generally for others. For example, if there is a reference placement for which $\mathfrak{S}_R(I) = 0$, I denoting the identity on $\mathcal{V}$, then the body in this placement is free of internal force-moments. In this case, the body is said to possess a *natural state*.

Material symmetries provide a second example of dependence on a particular reference placement. Let $\hat{\kappa} = (\varphi_S, \Phi_S)$ denote a second choice of the reference placement, with corresponding response function $\mathfrak{S}_S$. Then the mapping P carrying κ to $\hat{\kappa}$ and the deformations F for κ and G for $\hat{\kappa}$ are related through (2.2.29). A calculation parallel to that giving us $(2.2.30)_2$ shows that

$$\mathfrak{S}_S(G) = \mathfrak{S}_R(F)P^T, \qquad (2.7.4)$$

and this provides a general rule for the change in the response function of the material under a change of reference placement. For certain choices of $\hat{\kappa}$ it may happen that the response function is unchanged:

$$\mathfrak{S}_R = \mathfrak{S}_S. \qquad (2.7.5)$$

By (2.2.29) and (2.7.4) this requires that

$$\mathfrak{S}_R(F) = \mathfrak{S}_R(FP)P^T \qquad (2.7.6)$$

for all $F \in \mathscr{Gl}(\mathscr{V})$. Standard theory in continuum mechanics shows that the set of all tensors P satisfying (2.7.6) forms a group $\mathscr{G}$. We call it the *internal force-moment symmetry group* for the reference placement κ. If the material is hyperelastic, then invariance of the stored-energy function under a change in reference placement gives a parallel restriction:

$$\sigma(F) = \sigma(FP) \qquad (2.7.7)$$

for all $F \in \mathscr{Gl}(\mathscr{V})$. The set of all tensors P with this property forms a group $\mathscr{g}$ called the *energy symmetry group* for the reference placement κ. Any hyperelastic material has both an internal force-moment symmetry group $\mathscr{G}$ and an energy symmetry group $\mathscr{g}$. Generally, these are distinct, but certainly $\mathscr{g} \subset \mathscr{G}$.

The symmetry groups $\mathscr{G}$ and $\mathscr{g}$ depend on our choice of the reference placement.[18] An elastic or hyperelastic material is an *isotropic solid* if there is a reference placement κ for which the associated symmetry groups are $\mathscr{G} = \mathcal{O}(\mathscr{V})$ or $\mathscr{G} = \mathscr{g} = \mathcal{O}(\mathscr{V})$, respectively. In this case, κ is called an *undistorted state* of the body. It is characterized mathematically by the requirement that $\mathfrak{S}_R$ and σ satisfy the invariance conditions (2.7.6) and (2.7.7) for all $P \in \mathcal{O}(\mathscr{V})$.

One of the basic requirements of modern continuum mechanics is that the response function of a material be the same for all observers. Before stating this condition precisely, let us derive the rule of transformation for response functions in any change of frame. The general vector-valued response function $\mathfrak{s}_R$ is assumed to be frame indifferent. By the methods of Section 2(vi), this means that

$$\mathfrak{s}_R^*(F^*, N) = Q\mathfrak{s}_R(F, N), \qquad (2.7.8)$$

where F and F^* are related by $(2.6.6)_2$. Equivalently, in tensor form and for hyperelastic materials, we have

$$\mathfrak{S}_R^*(F^*) = Q\mathfrak{S}_R(F), \qquad \sigma^*(F^*) = \sigma(F). \qquad (2.7.9)$$

By eliminating F^* from each of these, we obtain explicit expressions for the response functions $\mathfrak{s}_R^*$, $\mathfrak{S}_R^*$, and σ^* to be used by any observer in the starred frame. The *principle of material frame indifference* states that all

---

[18] Symmetry groups, under the alternative name *peer group*, are described at great length by TRUESDELL [1977, §§IV.12–17]. In particular, he develops rules for the transformation of symmetry groups under a change of reference placement.

observers use the *same* response function:

$$\mathfrak{S}_R^* = \mathfrak{S}_R, \qquad \sigma^* = \sigma. \tag{2.7.10}$$

Combining this requirement with (2.7.9) and using $(2.6.6)_2$, we see that

$$\mathfrak{S}_R(QF) = Q\mathfrak{S}_R(F), \qquad \sigma(QF) = \sigma(F), \tag{2.7.11}$$

for all $Q \in \mathcal{O}(\mathscr{V})$ and all $F \in \mathscr{Gl}(\mathscr{V})$. In parallel to (2.7.6) and (2.7.7), these invariance relations restrict our choice of the response functions $\mathfrak{S}_R$ and σ; in contrast, however, they are assumed to apply regardless of our choice of the reference placement κ.

Representation theorems, now standard in continuum mechanics,[19] characterize all functions that satisfy invariance conditions such as (2.7.6), (2.7.7), and (2.7.11). Some initial reductions can be obtained directly from the polar decomposition

$$F = RU = VR, \tag{2.7.12}$$

where R is orthogonal, and U and V are called the right and left stretch tensors, respectively. Then the conditions of material frame indifference (2.7.11) impose on the internal force-moment tensors and the stored energy the reduced forms

$$\mathfrak{S}_R(F) = R\mathfrak{S}_u(U), \qquad \mathfrak{S}(F) = R\mathfrak{S}_u(U)UR^T, \qquad \sigma(F) = \sigma^u(U), \tag{2.7.13}$$

where $\mathfrak{S}_u$ and σ^u are arbitrary functions on the set of symmetric positive-definite tensors. For an isotropic material, (2.7.6) and (2.7.7) must hold for all $P \in \mathcal{O}(\mathscr{V})$, and these can be exploited in the same way to give the representations

$$\mathfrak{S}_R(F) = \mathfrak{S}_v(V)R, \qquad \mathfrak{S}(F) = \mathfrak{S}_v(V)V, \qquad \sigma(F) = \sigma^v(V), \tag{2.7.14}$$

in terms of new arbitrary functions $\mathfrak{S}_v$ and σ^v.

When isotropy and material frame indifference are imposed together, even more specialized representations can be obtained. For example, theorems on representations imply the forms

$$\mathfrak{S}(F) = Q_0(I, II, III)I + Q_1(I, II, III)FF^T + Q_2(I, II, III)(FF^T)^2,$$
$$\sigma(F) = \sigma^1(I, II, III), \tag{2.7.15}$$

where Q_0, Q_1, Q_2, and σ^1 are scalar functions of the *principal strain invariants* I, II, and III defined by

$$I \equiv F \cdot F = U \cdot U = V \cdot V,$$
$$II \equiv \tfrac{1}{2}\{I^2 - (FF^T) \cdot (FF^T)\}$$
$$= \tfrac{1}{2}\{I^2 - (U^2) \cdot (U^2)\} = \tfrac{1}{2}\{I^2 - (V^2) \cdot (V^2)\}, \tag{2.7.16}$$
$$III \equiv \det FF^T = \det U^2 = \det V^2.$$

---

[19] TRUESDELL and NOLL [1965, §§10–13].

If the material is an isotropic hyperelastic solid, then both representations (2.7.15) apply. Since $\mathfrak{S} = \mathfrak{S}_R F^T = (\partial \sigma / \partial F) F^T$, we find that the functions Q_0, Q_1, and Q_2 in this case are given in terms of the derivatives of σ^1 by

$$Q_0 = 2III \frac{\partial \sigma^1}{\partial III},$$

$$Q_1 = 2\left(\frac{\partial \sigma^1}{\partial I} + I \frac{\partial \sigma^1}{\partial II}\right), \qquad (2.7.17)$$

$$Q_2 = -2 \frac{\partial \sigma^1}{\partial II}.$$

An alternate way to account simultaneously for isotropy and material frame indifference is through the use of principal axes of strain and internal force-moment. Using (2.7.6) and (2.7.11)$_1$, we can show that the response function $\mathfrak{S}_u$ of (2.7.13) satisfies the invariance condition

$$\mathfrak{S}_u(QUQ^T) = Q\mathfrak{S}_u(U)Q^T \qquad (2.7.18)$$

for all $Q \in \mathcal{O}(\mathcal{V})$. Functions satisfying this condition are called *isotropic*. A theorem in linear algebra shows[20] that the principal axes of $\mathfrak{S}_u$ and U coincide. If a common orthonormal basis of these principal axes is denoted D_1, D_2, D_3, then by (2.7.13) we can write

$$\mathfrak{S}_R = \sum_{i=1}^{3} T_i R D_i \otimes D_i, \qquad U = \sum_{i=1}^{3} u_i D_i \otimes D_i. \qquad (2.7.19)$$

The T_i are the *principal force-moments* and the u_i are the *principal stretches*. In this formulation the T_i can be viewed as the response functions of the material. Their relation to the coefficients Q_0, Q_1, and Q_2 can be found by using (2.3.8)$_3$, (2.7.12), and (2.7.19) to express both sides of (2.7.15)$_1$ in terms of the principal axes. This gives the result

$$T_i = Q_0/u_i + Q_1 u_i + Q_2 u_i^3, \qquad i = 1, 2, 3. \qquad (2.7.20)$$

The representation (2.7.19)$_2$ for U, when placed in (2.7.16), allows us to express the principal invariants I, II, and III directly in terms of the principal stretches:

$$I = u_1^2 + u_2^2 + u_3^2,$$

$$II = u_1^2 u_2^2 + u_1^2 u_3^2 + u_2^2 u_3^2, \qquad (2.7.21)$$

$$III = u_1^2 u_2^2 u_3^2.$$

Each of these is symmetric in u_1, u_2, and u_3. As a result, any function of the principal invariants can also be viewed as a symmetric function of the principal stretches. For a hyperelastic material, this gives us the new representation

---

[20] Cf. TRUESDELL [1977, p. 167].

$$\sigma(F) = \sigma^s(u_1, u_2, u_3), \qquad\qquad (2.7.22)$$

where σ^s is symmetric in its three arguments. By placing (2.7.17) into (2.7.20) and expressing the derivatives of σ^1 with respect to I, II, and III in terms of derivatives of σ^s with respect to u_1, u_2, and u_3, we find that

$$T_i = \frac{\partial \sigma^s}{\partial u_i}, \qquad i = 1, 2, 3, \qquad\qquad (2.7.23)$$

a result that parallels the original relation $(2.7.3)_1$. We conclude[21] that there is a single function τ of the principal stretches, symmetric in its second two arguments, such that

$$T_i = \tau(u_i, u_j, u_k), \qquad i = 1, 2, 3, \qquad\qquad (2.7.24)$$

where i, j, and k are to be taken in cyclic order. Indeed, (2.7.23) shows that $\tau = \partial \sigma^s / \partial u_1$. As we shall see in Chapter 4, this formulation in terms of principal force-moments and principal stretches is particularly useful in the analysis of special problems.

A pseudo-rigid body may be subject to the constraint of *incompressibility*, which is defined by the kinematic condition

$$\det F = 1. \qquad\qquad (2.7.25)$$

Such a body exhibits no change of volume in its motion and so is unable to respond to a hydrostatic pressure. A pressure manifests itself in the theory here by equal double forces without moment in all directions and so, by the interpretations developed in Section 2(iv), is given by a current force-moment tensor M that is proportional to I. The response functions for incompressible isotropic elastic and hyperelastic materials contain terms that compensate for this pressure, and they have the form

$$\mathfrak{S}(F) = -pI + Q_1(I, II)FF^T + Q_2(I, II)(FF^T)^2,$$
$$\sigma(F) = -p \det F + \sigma^1(I, II). \qquad\qquad (2.7.26)$$

Q_1, Q_2, and σ^1 are now functions of only I and II, and p is an indeterminate scalar pressure. In specific applications, the equations of motion (2.3.9) together with the constraint (2.7.25) suffice to determine p as well as the deformation F as functions of time.

Certain specific stored-energy functions have proved useful in past studies of hyperelastic materials. One of the simplest is that for a *semilinear material* (JOHN [1960], LURE [1968]) as given by

$$\sigma = \sigma^s(u_1, u_2, u_3)$$
$$= \tfrac{1}{2}(\lambda + 2\mu)(u_1^2 + u_2^2 + u_3^2) - (3\lambda + 2\mu)(u_1 + u_2 + u_3)$$
$$+ \lambda(u_1 u_2 + u_1 u_3 + u_2 u_3) + 3(3\lambda + 2\mu)/2. \qquad (2.7.27)$$

---

[21] Cf. TRUESDELL and NOLL [1965, §48].

λ and μ are the Lamé coefficients,[22] which must satisfy the restrictions

$$\mu > 0, \qquad 3\lambda + 2\mu > 0. \tag{2.7.28}$$

Equivalent forms of the response function for this material are easily derived from (2.7.13), (2.7.19), (2.7.23), and (2.7.24), resulting in the expressions

$$T_i = \lambda(u_i + u_j + u_k - 3) + 2\mu(u_i - 1),$$

$$\tau(\xi, \eta, \zeta) = \lambda(\xi + \eta + \zeta - 3) + 2\mu(\xi - 1), \tag{2.7.29}$$

$$\mathfrak{S}_u = \lambda(\mathrm{tr}(U - I))I + 2\mu(U - I).$$

Each of these shows that the internal force-moment is linear in $U - I$ and vanishes when $U = I$. Indeed, this type of material is most useful in the analysis of small strains about a natural state of a body.

The stored-energy function of a *Hadamard–Green material* (SIMPSON and SPECTOR [1984]) is given by

$$\sigma = \sigma^1(I, II, III)$$

$$= \alpha_1 I + \alpha_2 II + f(III), \tag{2.7.30}$$

where α_1 and α_2 are constitutive constants and f is a function of III alone. If $f \equiv 0$, we obtain the special case called a *Mooney–Rivlin material*, and if further $\alpha_2 = 0$, the material is *neo-Hookean*. A *Blatz–Ko material* (BLATZ and KO [1962]) is one for which $\alpha_2 = 0$ and $f(III) = (III)^{-m}/m$ for some real number m. Equivalent forms of the constitutive relation for a general Hadamard–Green material are

$$\sigma^s = \alpha_1(u_1^2 + u_2^2 + u_3^2) + \alpha_2(u_1^2 u_2^2 + u_1^2 u_3^2 + u_2^2 u_3^2) + f(u_1^2 u_2^2 u_3^2),$$

$$T_i = 2[\alpha_1 + \alpha_2(u_j^2 + u_k^2) + u_j^2 u_k^2 f'(u_i^2 u_j^2 u_k^2)]u_i,$$

$$\tau(\xi, \eta, \zeta) = 2[\alpha_1 + \alpha_2(\eta^2 + \zeta^2) + \eta^2 \zeta^2 f'(\xi^2 \eta^2 \zeta^2)]\xi, \tag{2.7.31}$$

$$Q_0 = 2IIIf'(III), \qquad Q_1 = 2(\alpha_1 + \alpha_2 I), \qquad Q_2 = -2\alpha_2,$$

$$\mathfrak{S}_u = 2(\alpha_1 + \alpha_2 I)U - 2\alpha_2 U^3 + 2IIIf'(III)U^{-1}.$$

In order for the reference placement to be natural, $\mathfrak{S}_u$ must vanish when $U = I$. This requires the additional restriction that $\alpha_1 + 2\alpha_2 + f'(1) = 0$.

Finally, a *St. Vénant–Kirchhoff material* (TRUESDELL and NOLL [1965, Eq. (94.1)]) is defined by the stored-energy function

$$\sigma = \sigma^s(u_1, u_2, u_3)$$

$$= \frac{\lambda + 2\mu}{8}(u_1^4 + u_2^4 + u_3^4) + \frac{\lambda}{4}(u_1^2 u_2^2 + u_1^2 u_3^2 + u_2^2 u_3^2)$$

$$- \frac{3\lambda + 2\mu}{4}(u_1^2 + u_2^2 + u_3^2) + \frac{3(3\lambda + 2\mu)}{8}, \tag{2.7.32}$$

---

[22] We note that λ and μ have the dimensions of force-moment rather than those of stress.

where the coefficients λ and μ once again satisfy (2.7.28). As with the previous cases, we can express the constitutive relation for such a material in a variety of forms. Some of these are[23]

$$T_i = [\tfrac{1}{2}\lambda(u_i^2 + u_j^2 + u_k^2 - 3) + \mu(u_i^2 - 1)]u_i,$$

$$\tau(\xi, \eta, \zeta) = [\tfrac{1}{2}\lambda(\xi^2 + \eta^2 + \zeta^2 - 3) + \mu(\xi^2 - 1)]\xi,$$

$$Q_0 = 0, \qquad Q_1 = \tfrac{1}{2}\lambda(I - 3) - \mu, \qquad Q_2 = \mu,$$

$$\mathfrak{S}_u = [\tfrac{1}{2}\lambda(I - 3) - \mu]U + \mu U^3.$$

(2.7.33)

These show in particular that the internal force-moment vanishes when $U = I$, and so the reference placement for this material is a natural state. The St. Vénant–Kirchhoff material is perhaps the simplest exhibiting nonlinear response, and we shall use it extensively in Chapters 4 and 5 as a case study.

---

[23] The coefficients Q_0, Q_1, and Q_2 correct those reported in COHEN and MUNCASTER [1984, Eq. (3.20)]. This change affects only one of the solutions reported there, the new result being recorded here as our (4.2.48).

CHAPTER 3

Consistency with Other Continuum Theories

The theory of pseudo-rigid bodies represents a deformable body in terms of a single point moving in three-dimensional space and a tensor measuring changes in orientation and features of deformation. This is an extremely coarse description of a real body, especially when cast against the sophistication of most modern studies of deformable media. It is simply a reflection, however, of the class of motions that we choose to consider, and it must be viewed in relation to the tractability of the theory. We choose to consider only those motions of real bodies that are characterized largely by (1) a transplacement of the mass center, (2) a change of orientation similar to that appearing in the mechanics of rigid bodies, and (3) overall measures of extension–compression and shear. This admittedly limited description of motion must be weighed against the theory itself, in which the basic equations of motion, given by (2.3.9) or (2.3.13), form a system of ordinary differential equations. By restricting the class of motions of interest, we have obtained a theory much more tractable than ones based on initial-boundary-value problems for systems of partial differential equations.

The theory we have developed here stands quite independent of other continuum theories. With a basic knowledge of the system of forces acting on a body, we can now formulate equations of motion that describe how pseudo-rigid bodies move and deform in time. Methods of solution, analysis, and approximation of these equations will be our focus beginning in Chapter 4. However, the true value of all these results lies in our ability to map the coarse description we obtain into information about real bodies. This mapping is at present built from intuition and perceptions; the analyses of Chapters 4–6 will add experience and increase our building blocks to three. Most studies of directed continua go no further. In this chapter, we build such a mapping explicitly by considering in detail the relation between the theory of pseudo-rigid bodies and three-dimensional theories of continuum physics.

(i) The General Notion of Consistency

We wish to compare mathematically two theories of material response that exhibit an overlap in the classes of motions each is designed to address. One of these, the *coarse theory*, provides a gross, overall description of motion. The other, called the *fine theory*, is designed to provide very detailed information, much more detailed than we desire or need for the applications we have in mind. Since both theories address a common class of problems, we should expect their predictions in some sense to agree or to be *consistent*, at least for this class. In this chapter, we select nonlinear elastodynamics as the fine theory and we study the question of consistency between it and the theory of pseudo-rigid bodies. Other examples of fine–coarse theory pairs, and the historical motivations they provided for a general study of consistency, can be found in Section 1(i).

A detailed study of the question of consistency speaks to several aspects of the theory we have developed here. First, it provides a concrete way in which we may use motions predicted for a pseudo-rigid body to inform us about motions of truly three-dimensional bodies. More precisely, given any solution $(r(t), F(t))$ of the basic equations of motion (2.3.9), we map this solution into an associated motion, or a suitable approximate motion, for the equations of nonlinear elasticity. A second product of the study of consistency is the ability to make direct comparisons. Since the theory of nonlinear elasticity is well accepted in continuum physics, it serves as a familiar background against which we can appraise the value of the basic definitions, hypotheses, and principles of balance we have set down here. In particular, we shall see how the constitutive relation $(2.2.28)_2$ for the current moment of momentum tensor H and how at least a restricted version of the principle of material frame indifference (2.7.9) arise naturally out of the three-dimensional theory. The study of consistency also plays a role in the modeling of the external force f and the external force-moment M_R. The results (2.4.1) and (2.4.3) are appropriate only for a system of concentrated point forces. In order to obtain similar expressions that account for the effects of surface tractions and body forces, it is necessary to picture bodies as truly three-dimensional objects. Consistency is concerned precisely with this picture. In anticipation of the results we shall obtain here, we have already presented in (2.4.23) and (2.4.24) the forms of f and M_R for a continuous system of loads.

The *theory of consistency* developed here has two sides, one geometric and speculative in nature, the other analytic and rigorous. The geometric side provides a convenient pictorial representation of the ideas behind the analytical analysis, and it is useful to present it first. We use it to motivate and interpret the analysis that begins in Section 3(ii).

Nonlinear elasticity is concerned with a body that occupies in a reference placement κ some open three-dimensional region $\mathscr{B} \subset \mathscr{E}$. (Henceforth, as in Section 2(iv), we simplify notation by using the same symbol $\mathscr{B}$ to denote both the body and the region in the reference placement occupied by that body.) Places in κ are represented by the symbol X. A transplacement of the body is a mapping χ of these points into the places they occupy in the current placement:

$$x = \chi(t, X). \tag{3.1.1}$$

Thus, $x \in \mathscr{E}$ is the position at time t of the reference point $X \in \mathscr{B}$. The equations of motion of nonlinear elasticity, which we present in Section 3(ii), form a system of partial differential equations for determining the mapping χ. This mapping is the basic descriptor of the theory.

For a pseudo-rigid body, the basic descriptors are the current position $r(t)$ of the center of mass and the current value of the deformation tensor $F(t)$. At each time t the components of r, F, and their derivatives $s \equiv \dot{r}$ and $G \equiv \dot{F}$, identify a point in a space of dimension 24 called the *pseudo-rigid state space* $\mathscr{S}_p$. Each solution of the equations of motion (2.3.9) can be viewed as defining a curve in this state space. A similar interpretation applies to elasticity theory: the mappings χ and $\eta \equiv \dot{\chi}$ define for each time t a point in an infinite-dimensional *elastic state space* $\mathscr{S}_e$, a function space of mappings from $\mathscr{B}$ to $\mathscr{E} \times \mathscr{V}$. Each solution according to elasticity theory is a curve in this space. For the theory of pseudo-rigid bodies and the theory of nonlinear elasticity to be consistent, we require that there be a mapping that identifies with each solution curve in the pseudo-rigid state space a class of solution curves in the elastic state space. All the members of a given class are motions of an elastic body for which the associated pseudo-rigid motion provides some type of *asymptotic approximation*.

Geometrically, we introduce this mapping in two stages. In the first we assume that the *pseudo-rigid state space can be embedded*[1] *in the elastic state space* in the form of a 24-dimensional manifold $\mathscr{M}$. We call $\mathscr{M}$ the *coarsification* of the elastic state space. The embedding is called the *coarsifier* and is defined by a pair of mappings $(\mathbb{X}, \mathbb{N})$: $\mathscr{S}_p \to \mathscr{S}_e$. More precisely, to each pseudo-rigid state (r, s, F, G) there corresponds an elastic state (χ, η) given by

$$\chi(\cdot) = \mathbb{X}(\cdot; r, s, F, G),$$
$$\eta(\cdot) = \mathbb{N}(\cdot; r, s, F, G). \tag{3.1.2}$$

As a simple and explicit illustration of such mappings, consider

$$\chi(X) = x_0 + r + F(X - \varphi_R),$$
$$\eta(X) = s + G(X - \varphi_R), \tag{3.1.3}$$

---

[1] In this monograph we are concerned with local properties of the embedding. For analyses of global properties and their relationship to the general study of fine–coarse theory pairs, see PIERCE [1985].

in which φ_R denotes the position of the center of mass in the reference placement. In this case each pseudo-rigid state (r, s, F, G) is mapped into a homogeneous elastic state. The center of mass corresponding to this homogeneous deformation is at r while the constant deformation gradient is precisely F. Unfortunately, this illustration is misleading. While the manifold $\mathcal{M}$ might indeed have the simple form (3.1.3), we are not free to make such a choice. Rather, the manifold is to be determined by the further requirement that not only do pseudo-rigid states map into elastic states, but also pseudo-rigid motions map into elastic motions. That is, *given a solution $(r(t), F(t))$ according to the theory of pseudo-rigid bodies, if we set*

$$\chi(t, X) = \mathbb{X}(X; r(t), \dot{r}(t), F(t), \dot{F}(t)),$$

$$\eta(t, X) = \mathbb{N}(X; r(t), \dot{r}(t), F(t), \dot{F}(t)),$$

(3.1.4)

then χ must be a solution according to nonlinear elasticity and η must equal $\dot{\chi}$. This is a far reaching assumption. As we shall see in Section 3(ii), it is precisely this requirement that delivers a system of partial differential equations for the functions $\mathbb{X}$ and $\mathbb{N}$ that define the manifold. In the language of differential equations, (3.1.4) states that $\mathcal{M}$ must be an *invariant manifold* for the equations of motion of nonlinear elasticity: any elastic solution curve that at one time lies on $\mathcal{M}$ must lie on that manifold at all future times.

The elastic motions defined by (3.1.4), being parameterized exactly by the set of *all* motions of a pseudo-rigid body, make up only a small subset of all the solutions that elasticity theory delivers. As the second stage in our construction, we focus on *the asymptotic status of the manifold $\mathcal{M}$ with respect to all other solutions in nonlinear elasticity that lie near it.* If $\mathcal{M}$ is asymptotically stable in time, then we may expect that any elastic solution curve that begins in a neighborhood of the manifold will approach as $t \to \infty$ some solution on the manifold, that is, the exact image of some pseudo-rigid motion. If $\mathcal{M}$ is only stable, then each elastic solution initially near the image of a pseudo-rigid motion should continue to lie near it. The precise notion of "asymptotic" could be of a quite different type, referring not to limiting behavior in time but rather as a parameter, such as the total mass m or the diameter of the body $\mathcal{B}$, approaches some specified value. In any case, this asymptotic status allows us to match general solutions with those on the coarsification $\mathcal{M}$, completely within the context of the theory of nonlinear elasticity.

This geometric side of the concept of consistency is summarized in Fig. 1. Each pseudo-rigid motion, viewed as a curve in the pseudo-rigid state space $\mathcal{S}_p$, is mapped into a special exact solution according to nonlinear elasticity. The set of all these special solutions forms the coarsification of elasticity theory, a 24-dimensional manifold $\mathcal{M}$ embedded in the elastic state space $\mathcal{S}_e$. Finally, each general solution in the theory of elasticity is related, in some asymptotic sense, to the image of a pseudo-

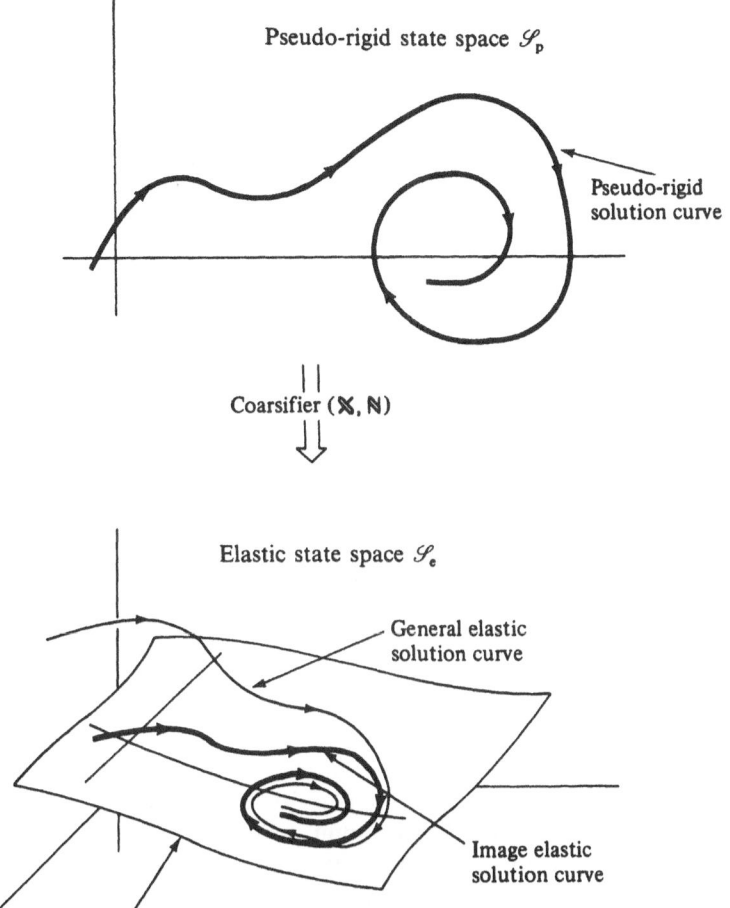

Figure 1

rigid motion, as represented here by the image elastic solution curve on the manifold.

This picture is not a summary of rigorously proved results. It is a suggestion of how motions of pseudo-rigid bodies might be related to motions of elastic bodies in order that the corresponding theories exhibit a sense of consistency. The analytic side of the study of consistency is concerned with giving rigorous justification to the various parts of this geometric picture. It includes the following main issues, suggested in part by the geometry:

(1) proof of the existence of an embedded invariant manifold,
(2) analysis of the uniqueness of the manifold,

(3) derivation and proof of the asymptotic status of the manifold in relation to neighboring elastic solutions,

(4) analysis of properties of the manifold, such as governing equations and symmetry features,

(5) approximation of the manifold.

Several of these issues, and in particular those concerning existence, uniqueness, and asymptotic behavior, present challenging open problems in analysis.[2] In this short monograph, beginning in Section 3(ii), we examine the more modest problems presented by issues (4) and (5). That is, assuming that a special manifold of elastic solutions exists, one that can be viewed as an embedding in nonlinear elasticity of the full class of pseudo-rigid motions, we examine the properties of this manifold and ways in which the manifold might be approximated.

(ii) Elasticity Theory. Subtheories

The material comprising a three-dimensional body $\mathscr{B}$ is *elastic* if its stress is given in terms of the deformation χ by a constitutive relation of the form

$$T_R = \mathfrak{h}_R(\nabla\chi). \tag{3.2.1}$$

T_R is the first Piola–Kirchhoff stress tensor, and $\mathfrak{h}_R$ is the response function of the material. Let the external forces acting on the body consist of a field of body force $b(x)$ per unit mass distributed over the body and a field of surface tractions $t(x)$ per unit area in the reference placement distributed over its boundary. Then the boundary-value problem posed by EULER's balance of linear momentum is

$$
\begin{aligned}
\rho_R \ddot{\chi} &= \operatorname{div} \mathfrak{h}_R(\nabla\chi) + \rho_R b(\chi), & X \in \mathscr{B}, \\
\mathfrak{h}_R(\nabla\chi)N &= t(\chi), & X \in \partial\mathscr{B}.
\end{aligned}
\tag{3.2.2}
$$

---

[2] The geometric picture we have presented here is based on more than idle speculation. The central idea is that the asymptotic forms of general solutions of a problem are often special solutions of that same problem; moreover, these special solutions collectively define an invariant set. In the context of mechanics, TRUESDELL and MUNCASTER [1980, Chapters XIII, XIV, XV] present extensive illustrations of this fact in the kinetic theory of gases, where they introduce the term *principal solution*. In mathematics more generally, this same idea is at the very roots of much of modern day spectral theory for linear differential equations. For nonlinear problems in mathematics, the *theory of invariant manifolds* provides an important body of analytic results that addresses precisely an abstract analogue of the geometric picture we have given here. This includes proofs of existence and asymptotic status of the invariant manifold and discussions of the issue of uniqueness. These analyses should serve as prototypes for the problems raised here in (1), (2), and (3). For a review of this theory, and its possible bearing on the rigorous study of consistency, see MUNCASTER [1984a, §§5, 6].

ρ_R is the mass density in the reference placement, and $N(X)$ is the unit outward normal to $\partial\mathscr{B}$ at X. An over dot denotes a partial derivative with respect to time, and ∇ and div are calculated with respect to the reference place X. Each of ρ_R, b, t, and the response function $\mathfrak{h}_R$ may depend explicitly on X, but we suppress this dependence since it plays a nonessential role.

Particular elastic materials are characterized through special properties of the response function $\mathfrak{h}_R$. If $\mathfrak{h}_R(X, \nabla\chi)$ does not depend explicitly on the reference place X, the material is *homogeneous*; if in addition $\mathfrak{h}_R = 0$ when $\nabla\chi = I$, then the reference placement is a *natural state* for the material. Material symmetry can also be introduced through invariance conditions analogous to (2.7.6). In parallel to the current internal force-moment $\mathfrak{S}$, we may define an elastic material also in terms of the *Cauchy stress* T, which measures the stress relative to the current placement of the body. It is specified by a constitutive relation $T = \mathfrak{h}(\nabla\chi)$, and the response functions $\mathfrak{h}_R$ and $\mathfrak{h}$ are related by the transformation $\mathfrak{h} = \mathfrak{h}_R F^T/J$, where $J = \det F$ (cf. $(2.3.8)_3$).

The equations of motion (3.2.2), together with initial values for χ and $\dot\chi$, form the fine theory of interest. If we set

$$\eta = \dot\chi, \tag{3.2.3}$$

then the set of pairs of fields (χ, η) makes up the elastic state space $\mathscr{S}_e$.

Geometrically, we wish to view the set of solutions of the equations of motion for pseudo-rigid bodies as an invariant manifold $\mathscr{M}$ in the state space for this fine theory. The analytic approach here supports this view but begins from a more basic and physical notion, that of a *subtheory*. This notion plays an important role in the study of consistency in general, and so initially it is useful to examine it independently. Critical to the notion of a subtheory is the assumption that the equations of the parent theory be *autonomous*. This is the reason that we have assumed that b, t, $\mathfrak{h}_R$, etc., do not depend explicitly on the time t.

Consider the class of initial values

$$\chi|_{t=0} = \chi_0, \qquad \dot\chi|_{t=0} = \eta|_{t=0} = \eta_0, \tag{3.2.4}$$

for which (3.2.2) has a unique solution[3] in some interval of time about $t = 0$. Suppressing the dependence of χ, χ_0, η_0, etc., on X, we write this solution as follows as a function χ_{sol} of the initial values:

$$\chi(t) = \chi_{sol}(t; \chi_0, \eta_0). \tag{3.2.5}$$

The *uniqueness of solutions* imposes on χ_{sol} the identity

$$\chi_{sol}(t + \tau; \chi_0, \eta_0) = \chi_{sol}(t; \chi_{sol}(\tau; \chi_0, \eta_0), \dot\chi_{sol}(\tau; \chi_0, \eta_0)). \tag{3.2.6}$$

Our emphasis on uniqueness is a reflection of the way in which we

---

[3] See HUGHES, KATO, and MARSDEN [1977].

choose to study a theory. A "study" consists broadly in comparing one solution with others, and it is important to have a means of distinguishing between solutions, that is, a parameterization of the solutions of interest. Since dynamic behavior is our main concern, we parameterize here in terms of initial values and so focus only on the class of solutions for which this type of parameterization is possible.[4]

It is not necessary in studying a theory to consider all its solutions simultaneously. We may be interested, instead, in only a restricted class of solutions, represented parametrically as a set

$$\mathscr{F} = \{\chi_p \colon \mathscr{I}_p \to \mathscr{E} \,|\, p \in \mathscr{P}\}. \tag{3.2.7}$$

$\mathscr{P}$ is an open set is some space of parameters, and, for each admissible parameter value p, $\mathscr{I}_p$ is an open interval of times about $t = 0$ on which the solution χ_p is defined. As the following examples show, the size of the family $\mathscr{F}$ is determined by the size of the parameter set $\mathscr{P}$.

EXAMPLE 1 (A class of general solutions). Let $\mathscr{P}$ be an open subset of the elastic state space $\mathscr{S}_e$. Then p is a field pair (χ_0, η_0), which we may regard as a choice of initial values for a solution of (3.2.2). In this case the family $\mathscr{F}$ is given explicitly in terms of χ_{sol}:

$$\mathscr{F} = \{\chi_{sol}(\cdot\,; p)\,|\, p \in \mathscr{P} \subset \mathscr{S}_e\}. \tag{3.2.8}$$

This is a class of general solutions of (3.2.2).

EXAMPLE 2 (A static solution). Let χ_s be any given static solution of (3.2.2). Then

$$\mathscr{F} = \{\chi_s\}, \tag{3.2.9}$$

a singleton family. In this case, since there is only one solution of interest, we have omitted the singleton parameter set $\mathscr{P}$. Such a family would be important if we were principally interested in static deformations of the body under the action of a given set of surface loads. The first example represents one extreme in the size of $\mathscr{F}$, namely, a class of general solutions, while this example represents the other extreme, a single solution.

EXAMPLE 3 (Semi-inverse solutions). Classes of semi-inverse solutions represent examples that lie between the above extremes. As an illustration, consider the quasistatic homogeneous deformation

$$\chi(t, X) = F(t)(X - X_0) + r(t) + x_0, \tag{3.2.10}$$

where X_0 is a constant. Setting $b = 0$ and regarding $(3.2.2)_2$ as a formula for the field of tractions that must be applied to produce such a defor-

---

[4] In contrast, for static experiments on buckling, solutions are often parameterized in terms of loading parameters and geometric features of the buckled state.

mation, we see that (3.2.2) reduces to

$$\ddot{r} = 0, \qquad \ddot{F} = 0. \tag{3.2.11}$$

The initial conditions for this system define a parameter p and the corresponding set $\mathscr{P}$:

$$p = (r(0), \dot{r}(0), F(0), \dot{F}(0)),$$
$$\mathscr{P} = \mathscr{V} \times \mathscr{V} \times \mathscr{Gl}(\mathscr{V}) \times (\mathscr{V} \otimes \mathscr{V}). \tag{3.2.12}$$

With these preparations, we can describe the solutions of this semi-inverse family as the set

$$\mathscr{F} = \{F_p(X - X_0) + r_p + x_0 | p \in \mathscr{P}\}, \tag{3.2.13}$$

where r_p and F_p are the unique solutions of (3.2.11) and (3.2.12).

Uniqueness implies that a given set of solutions of (3.2.2) is characterized completely by the set of initial values of its members. These initial values define a mapping $(\mathbb{X}, \mathbb{N}): \mathscr{P} \to \mathscr{S}_e$ by

$$\mathbb{X}(p) = \chi_p(0), \qquad \mathbb{N}(p) = \eta_p(0). \tag{3.2.14}$$

For example, for the semi-inverse solutions (3.2.13), let us write the parameter p generally as

$$p = (r, s, F, G). \tag{3.2.15}$$

Then for this case (3.2.10), (3.2.11), and (3.2.12) show that

$$\mathbb{X}(r, s, F, G) = F(X - X_0) + r + x_0,$$
$$\mathbb{N}(r, s, F, G) = G(X - X_0) + s. \tag{3.2.16}$$

Returning to the general family $\mathscr{F}$ given by (3.2.7), we see by (3.2.5) and (3.2.14) that its members are the solutions

$$\chi_p(t) = \chi_{\text{sol}}(t; \mathbb{X}(p), \mathbb{N}(p)). \tag{3.2.17}$$

Therefore, the initial values determining $\mathscr{F}$ define the set

$$\mathscr{M} = \{(\mathbb{X}(p), \mathbb{N}(p)) | p \in \mathscr{P}\}. \tag{3.2.18}$$

Each of the explicit sets $\mathscr{F}$ presented above is a particular example of a subtheory of nonlinear elasticity. However, there is a significant difference between the way in which these examples are constructed and the way in which we use subtheories in the study of consistency. In each of the examples the set $\mathscr{M}$ is prescribed *a priori*, generally through an explicit assumption about the form of the solutions in $\mathscr{F}$. The members of the class of general solutions (3.2.8) are values of χ_{sol}; the semi-inverse solutions (3.2.13) are quasistatic deformations. Given $\mathscr{M}$, we can proceed immediately to an analysis of the members of $\mathscr{F}$. For the class of semi-inverse solutions, this leads us to the system (3.2.11).

In the study of consistency we turn this procedure around: given that $\mathscr{F}$ is jointly characterized by a set of differential equations and a parameter set $\mathscr{P}$, can we find $\mathscr{M}$? Of course, given $\mathscr{M}$ we can then use (3.2.17) to find the members of $\mathscr{F}$ explicitly. Within the application of interest here, we can restate this idea more explicitly: beginning with the theory of pseudo-rigid bodies, and choosing as a parameter space the set $\mathscr{P}$ of initial values for the differential equations of that theory, can we find a manifold $\mathscr{M}$ in the elastic state space which defines the set of initial values for a subtheory $\mathscr{F}$ of nonlinear elasticity? The previous example of the semi-inverse solutions offers a suggestion as to how we might approach this problem. The set of semi-inverse solutions (3.2.13) exhibits an important sense of *closure*: any solution, which at one time has the form of a homogeneous deformation, will continue to have this form at all other times. Therefore, we can study these solutions purely within the space of homogeneous states. In a sense then, semi-inverse solutions comprise within themselves a separate theory, and our use of the term subtheory is a reflection of this fact.

For the general analysis of consistency, we need an analytic characterization of this notion of closure. The definition we adopt here is strongly linked to the way in which we choose to study solutions. We are interested in dynamic behavior. We study this behavior by recording instantaneous views, or snapshots, of the body as it moves. If these pictures are all we can record, then to obtain closure of our study we must require that each snapshot be an admissible initial state for the class of motions we are examining. More precisely, beginning with one solution $\chi_p \in \mathscr{F}$, we observe the state $(\chi_p(\tau), \eta_p(\tau))$ of this solution at any time τ, and then we require that this state be an admissible initial value for some other solution in the family. This means that the family, through observations in the past or the future, reproduces itself. Since such families exhibit a sense of closure more characteristic of a separate, independent theory, we formalize them through the

Definition. *A family* $\mathscr{F} = \{\chi_p \colon \mathscr{I}_p \to \mathscr{E} \,|\, p \in \mathscr{P}\}$ *of solutions from nonlinear elasticity is a* subtheory *if*

$$\chi_{\text{sol}}(\cdot\,; \chi_p(\tau), \eta_p(\tau)) \in \mathscr{F} \tag{3.2.19}$$

for all $\tau \in \mathscr{I}_p$ *and for all* $p \in \mathscr{P}$.

Each of our previous examples defines a subtheory in this sense. For the class of general solutions in Example 1, the snapshot $\chi_{\text{sol}}(\tau; p)$ lies in the open set $\mathscr{P} \subset \mathscr{S}_\text{e}$ for all τ sufficiently small. Therefore, it qualifies as an initial value for another solution in $\mathscr{F}$. The static solution of Example 2 is trivially a subtheory[5] since it does not change in time.

---

[5] Of course, if one is indeed interested only in static deformations, then a parametrization of solutions in terms of initial values is rather unnatural (cf. footnote 4).

Classes of semi-inverse solutions, as we have noted earlier, naturally form subtheories. Indeed, it is an underlying assumption in any application of the semi-inverse method that the special form chosen for solutions is consistent with (3.2.2). That is, any solution of (3.2.2) initially having a particular semi-inverse form will continue to have that same form at all future times.

There is an alternate form of (3.2.19) that is more useful for analysis. Recall that each of the solutions in $\mathcal{F}$ is determined by a single parameter value $p \in \mathcal{P}$. Thus, (3.2.19) states that there is a parameter value $P_p(\tau) \in \mathcal{P}$, depending on both p and τ, such that

$$\chi_{P_p(\tau)}(t) = \chi_{\text{sol}}(t; \chi_p(\tau), \eta_p(\tau)). \qquad (3.2.20)$$

In view of (3.2.6), (3.2.14), and (3.2.17), we can write this equivalently as

$$\begin{aligned} \chi_{P_p(\tau)}(t) &= \chi_p(t + \tau), \\ \eta_{P_p(\tau)}(t) &= \eta_p(t + \tau), \end{aligned} \qquad (3.2.21)$$

where, in view of (3.2.3), the second is the derivative of the first with respect to t. If we set $t = 0$, replace τ by t, and then use (3.2.14), we obtain

$$\begin{aligned} \chi_p(t) &= \mathbb{X}(P_p(t)), \\ \eta_p(t) &= \mathbb{N}(P_p(t)). \end{aligned} \qquad (3.2.22)$$

For the semi-inverse solutions of Example 3, (3.2.16) shows that the solutions given in (3.2.13) have precisely this form. In general, (3.2.22) states that each of the solutions in the subtheory is determined by (1) a curve in the parameter set $\mathcal{P}$, as defined by the function $P_p(\cdot)$, and (2) an embedding of this curve in the state space of the fine theory, as defined by the pair of mappings $(\mathbb{X}, \mathbb{N})$. This agrees with the geometric view of consistency outlined in Section 3(i). Alternately, *a necessary condition that a family $\mathcal{F}$ be a subtheory is that it define an invariant manifold $\mathcal{M}$ in the state space of the associated fine theory.*

(iii) The Subtheory of Pseudo-rigid Motions

The foregoing remarks concern the structure of subtheories in general. Henceforth, we focus exclusively on the theory of pseudo-rigid bodies and nonlinear elasticity and examine in detail the characterization of subtheories interrelating these two. We begin with two preliminary steps. First, the basic variables of the theory of pseudo-rigid bodies, namely, m, E_R, r, and F, are identified with corresponding gross features of variables from the theory of elasticity. In the second, we select a set $\mathcal{P}$ through which the class of pseudo-rigid motions can be parameterized.

The *total mass m* and the *referential Euler tensor* E_R are defined by[6]

$$m = \int_{\mathscr{B}} \rho_R,$$

$$E_R = \int_{\mathscr{B}} \rho_R(X - \varphi_R) \otimes (X - \varphi_R),$$

(3.3.1)

where we recall that all volume and surface measures are suppressed. Since φ_R is the location of the center of mass in κ, we have

$$\int_{\mathscr{B}} \rho_R(X - \varphi_R) = \mathbf{0}.$$

(3.3.2)

In order to represent through r and F certain average features of the elastic deformation χ, we focus attention on the zeroth and first tensor moments of χ with respect to the displacement $X - \varphi_R$ from the center of mass. Normalizing these moments with respect to measures of mass, we define the *current center of mass* $r(t)$ and the *deformation* $F(t)$ as follows:

$$r(t) = \frac{1}{m} \int_{\mathscr{B}} \rho_R(\chi(t, X) - x_0),$$

$$F(t) = \int_{\mathscr{B}} \rho_R(\chi(t, X) - x_0) \otimes (X - \varphi_R)E_R^{-1},$$

(3.3.3)

where x_0 denotes a fixed origin in the current placement. These have the convenient property that if $\chi(t, X) = X$ for all t, so there is no deformation of the body from its position in the reference placement, then $r(t)$ is the displacement $\varphi_R - x_0$ of the center of mass relative to x_0 and $F(t) = I$. This indicates that F plays a role for pseudo-rigid bodies analogous to the role of the deformation gradient $\nabla\chi$ for three-dimensional bodies.

The center of mass r, being a basic variable in the mechanics of rigid bodies, is a natural choice for one of the descriptors of a pseudo-rigid body. In contrast, similar motivation for $(3.3.3)_2$ is less obvious. We note first that if χ is the rigid motion

$$\chi(t, X) = x_0 + r(t) + R(t)(X - \varphi_R),$$

(3.3.4)

$R(t)$ being an orthogonal tensor, then (3.3.1) and (3.3.2) imply that

$$F(t) = \left[r(t) \otimes \int_{\mathscr{B}} \rho_R(X - \varphi_R) + R(t) \int_{\mathscr{B}} \rho_R(X - \varphi_R) \otimes (X - \varphi_R) \right] E_R^{-1},$$

$$= R(t).$$

(3.3.5)

Hence, F reduces for a rigid motion to the orthogonal transformation R, precisely the second variable that appears in rigid-body mechanics. This fact generalizes our observation that $F = I$ when there is no deformation.

---

[6] TRUESDELL [1977, p. 44].

In precisely the same manner we can show that *if the motion of a three-dimensional body is affine*, that is, the sum of a translation and a homogeneous deformation, *then* (3.3.3) *gives this motion the form*

$$\chi(t, X) = x_0 + r(t) + F(t)(X - \varphi_R). \tag{3.3.6}$$

Our definition of F is a special case of the general weighted average

$$F(t) = \int_{\mathscr{B}} \rho_R(\chi(t, X) - x_0) \otimes \psi(X) E_\psi^{-1}, \tag{3.3.7}$$

where

$$E_\psi = \int_{\mathscr{B}} \rho_R \psi(X) \otimes \psi(X). \tag{3.3.8}$$

ψ is any vector-valued function of X whose mass average is null:

$$\int_{\mathscr{B}} \rho_R \psi(X) = 0. \tag{3.3.9}$$

This generalization is useful, for example, when an application suggests the use of a special set of orthogonal functions $\psi^{(0)}, \psi^{(1)}, \psi^{(2)}, \ldots$. These functions might provide special boundary values or be otherwise tailored to the given problem. Let $\psi^{(0)}(X) = 1$, $\psi^{(1)}(X) = \psi(X)$, and assume that generally $\psi^{(k)}(X)$ is a k-tensor-valued function of X satisfying the orthogonality relations

$$\int_{\mathscr{B}} \rho_R \psi^{(k)} \otimes \psi^{(l)} = 0 \quad \text{for } k \neq l. \tag{3.3.10}$$

Note that (3.3.2) is the special case of this relation when $k = 0$, $l = 1$, and $\psi(X) = X - \varphi_R$. By repeating the steps that led to (3.3.6), we find that the most general motion that is linear in $\psi^{(0)}$ and $\psi^{(1)}$ is given by

$$\chi(t, X) = x_0 + r(t) + F(t)\psi(X), \tag{3.3.11}$$

where F and r are defined by (3.3.7) and (3.3.3)$_1$, respectively.

Our objective is to find a subtheory of nonlinear elasticity that represents the class of pseudo-rigid motions. It is natural to parameterize the members χ_p of this subtheory in terms of the initial values of a general pseudo-rigid motion. In terms of the transplacement $(r(t), F(t))$, a typical value of the parameter p is the four-tuple $(r(0), \dot{r}(0), F(0), \dot{F}(0))$. For simplicity we write p in the form

$$p = (r, s, F, G) \tag{3.3.12}$$

and let $(r_p(t), F_p(t))$ denote the transplacement determined by p. Thus, r, s, F, and G are the initial values of $r_p(t)$, $\dot{r}_p(t)$, $F_p(t)$, and $\dot{F}_p(t)$, respectively. Moreover, we choose the parameter set $\mathscr{P}$ to be an open set such that

$$\mathscr{P} \subset \mathscr{V} \times \mathscr{V} \times \mathscr{Gl}(\mathscr{V}) \times (\mathscr{V} \otimes \mathscr{V}) = \mathscr{S}_p. \tag{3.3.13}$$

Setting $t = 0$ in (3.3.3), we obtain r and F explicitly in terms of $\chi_p(0)$. Similarly, by differentiating (3.3.3) with respect to time and setting $t = 0$, we obtain s and G in terms of $\eta_p(0) = \dot{\chi}_p(0)$. These formulas provide a one-to-one mapping between the parameter values p and the functions χ_p:

$$r = \frac{1}{m} \int_{\mathcal{B}} \rho_R(\chi_p(0) - x_0),$$

$$s = \frac{1}{m} \int_{\mathcal{B}} \rho_R \eta_p(0),$$

$$F = \int_{\mathcal{B}} \rho_R(\chi_p(0) - x_0) \otimes (X - \varphi_R)E_R^{-1}, \qquad (3.3.14)$$

$$G = \int_{\mathcal{B}} \rho_R \eta_p(0) \otimes (X - \varphi_R)E_R^{-1}.$$

Thus, r, s, F, and G are the coarse descriptors associated with the initial state $(\chi_p(0), \eta_p(0))$ of the elastic deformation χ_p.

So far we have considered purely matters of identification. How do we identify with a general elastic state (χ, η) some corresponding pseudo-rigid state (r, s, F, G)? What is the one-to-one correspondence between parameter values p and the members χ_p of a family of elastic motions? The answers to these questions made no use of the concept of a sub-theory; they are valid for any functions $\chi_p: \mathcal{I}_p \to \mathcal{E}$. Let us consider now the restrictions that subtheories impose. *We assume henceforth that the family of solutions χ_p forms a subtheory of nonlinear elasticity.* By the characterization (3.2.22), this means that there is a pair of mappings $(\mathbb{X}, \mathbb{N}): \mathcal{P} \to \mathcal{S}_e$, called the *coarsifier*, such that each solution χ_p is the image in the elastic state space $\mathcal{S}_e$ of a curve $t \mapsto (r_p(t), s_p(t), F_p(t), G_p(t))$ in the parameter set $\mathcal{P}$. More precisely,

$$\chi_p(t) = \mathbb{X}(r_p(t), s_p(t), F_p(t), G_p(t)),$$

$$\eta_p(t) = \mathbb{N}(r_p(t), s_p(t), F_p(t), G_p(t)). \qquad (3.3.15)$$

There is an important similarity between (3.3.15) and the semi-inverse assumption (3.2.10). Both are statements about a special form that is being imposed on the function χ. Each states that χ is determined completely by certain time-dependent vectors and tensors, that is, by $F(t)$ and $r(t)$ in the case of (3.2.10), and by $r_p(t)$, $s_p(t)$, $F_p(t)$, and $G_p(t)$ for (3.3.15). In this sense, both are semi-inverse assumptions. The difference between them lies in the form of the function that maps these vectors and tensors into χ. In (3.2.10) this function is explicit, namely, an affine transformation, but for (3.3.15) it is given by the unknown mappings $\mathbb{X}$ and $\mathbb{N}$. The choice of an affine transformation was based on informed speculation, for this is the essence of the semi-inverse method. Given this choice, we were left only to determine the functions $F(t)$ and $r(t)$. For (3.3.15), in contrast, we have replaced informed speculation with the precision of the closure property (3.2.19). As a result, we must find not only $r_p(t)$, $s_p(t)$, $F_p(t)$, and

$G_p(t)$ but also the functions $\mathbb{X}$ and $\mathbb{N}$. In this sense, *the search for subtheories is equivalent to a search for semi-inverse hypotheses.*

The requirement that the functions χ_p form a subtheory consists of two parts. One is the closure property as summarized in (3.3.15), and we examine its implications first. The second is the assumption that for all $p \in \mathscr{P}$ the function χ_p satisfies the initial-boundary-value problem (3.2.2) of nonlinear elasticity. As we show now, together these give rise to (1) a functional differential system for $\mathbb{X}$ and $\mathbb{N}$, and (2) a system of differential equations for r_p, s_p, F_p, and G_p formally equivalent to the equations of motion from the theory of pseudo-rigid bodies.

Recall from (3.2.14) that $\mathbb{X}$ and $\mathbb{N}$ are the initial values of χ_p and η_p. This implies that (3.3.14) can be rewritten as

$$r = \frac{1}{m} \int_{\mathscr{B}} \rho_R(\mathbb{X}(r, s, F, G) - x_0),$$

$$s = \frac{1}{m} \int_{\mathscr{B}} \rho_R \mathbb{N}(r, s, F, G),$$

$$F = \int_{\mathscr{B}} \rho_R(\mathbb{X}(r, s, F, G) - x_0) \otimes (X - \varphi_R)E_R^{-1},$$

$$G = \int_{\mathscr{B}} \rho_R \mathbb{N}(r, s, F, G) \otimes (X - \varphi_R)E_R^{-1}.$$

(3.3.16)

These are to hold as identities in the parameter $p = (r, s, F, G)$ and so provide restrictions on $\mathbb{X}$ and $\mathbb{N}$. If we replace r, s, F, and G by $r_p(t)$, $s_p(t)$, $F_p(t)$, and $G_p(t)$, respectively, we see from (3.3.15)$_1$ and (3.3.16)$_{1,3}$ that r_p and F_p are determined by χ_p according to our original definitions (3.3.3). Moreover, by (3.2.3), (3.3.15), and (3.3.16)$_{2,4}$, we have

$$\dot{r}_p = s_p, \qquad \dot{F}_p = G_p. \tag{3.3.17}$$

Finally, (3.3.14) shows that the initial conditions

$$r_p(0) = r, \qquad s_p(0) = s,$$
$$F_p(0) = F, \qquad G_p(0) = G,$$

(3.3.18)

apply.

Let us turn now to the initial-boundary-value problem of nonlinear elasticity. As an initial step we take tensor moments of (3.2.2)$_1$ analogous to those that define r and F. By integrating (3.2.2)$_1$ over $\mathscr{B}$ and applying the divergence theorem and the boundary conditions, we obtain from (3.3.15) the equation of balance

$$\dot{p} = m\ddot{r} = f, \tag{3.3.19}$$

where

$$f = \mathfrak{f}(r_p, s_p, F_p, G_p) \equiv \mathbb{F}(\mathbb{X}(r_p, s_p, F_p, G_p)). \tag{3.3.20}$$

$\mathbb{F}$ is the *net external force functional*

$$\mathbb{F}(\chi) = \int_{\partial\mathcal{B}} t(\chi) + \int_{\mathcal{B}} \rho_R b(\chi). \tag{3.3.21}$$

Similarly, by multiplying $(3.2.2)_1$ by $\otimes (X - \varphi_R)$, integrating over $\mathcal{B}$, and using the divergence theorem and the boundary conditions again, we obtain from (3.3.15) the equation of balance

$$\dot{H}_R = \dot{F}E_R = M_R - \Sigma_R, \tag{3.3.22}$$

where

$$M_R = \mathfrak{M}(r_p, s_p, F_p, G_p) \equiv \mathbb{M}(\mathbb{X}(r_p, s_p, F_p, G_p)) \tag{3.3.23}$$

and

$$\Sigma_R = \mathfrak{S}_R(r_p, s_p, F_p, G_p) \equiv \mathbb{S}(\mathbb{X}(r_p, s_p, F_p, G_p)). \tag{3.3.24}$$

$\mathbb{M}$ is the *external force-moment functional*

$$\mathbb{M}(\chi) = \int_{\partial\mathcal{B}} t(\chi) \otimes (X - \varphi_R) + \int_{\mathcal{B}} \rho_R b(\chi) \otimes (X - \varphi_R), \tag{3.3.25}$$

and $\mathbb{S}$ is the *net stress functional*

$$\mathbb{S}(\chi) = \int_{\mathcal{B}} \mathfrak{h}_R(\nabla\chi). \tag{3.3.26}$$

We see in (3.3.19) and (3.3.22) the basic equations of balance of the theory of pseudo-rigid bodies. Two particular features of these equations are important. First, the external force f and the referential external force-moment M_R are given in terms of the surface tractions and the body force by (3.3.20) and (3.3.23). When these are written explicitly, by use of (3.3.21) and (3.3.25), as integrals over the body and its boundary, we formally obtain the expressions (2.4.23) and $(2.4.24)_1$ hinted at in Section 2(iv). The parallel is not exact: the variable x in (2.4.23) and (2.4.24) is $\varphi + F(X - \varphi_R)$ while here it is $\mathbb{X}(r_p, s_p, F_p, G_p)$. We shall resolve this difference in Section 3(v) when we examine approximations for $\mathbb{X}$. A second, and perhaps more profound, observation is that (3.3.24) provides a constitutive relation, induced by nonlinear elasticity, for the referential internal force-moment tensor Σ_R. This relation between response functions will be examined in depth in Section 3(iv). The fact that r_p, s_p, and G_p are arguments of $\mathfrak{S}_R$ suggests a theory of pseudo-rigid bodies more general than that based on (2.7.1), but on this point too we shall have more to say in Section 3(v).

At this stage we have explored only certain moments of the system (3.2.2). Now we substitute χ_p directly. By the chain rule,

$$\dot{\chi}_p = \partial_r \mathbb{X} \dot{r}_p + \partial_s \mathbb{X} \dot{s}_p + \partial_F \mathbb{X} \dot{F}_p + \partial_G \mathbb{X} \dot{G}_p,$$

$$\ddot{\chi}_p = \dot{\eta}_p \tag{3.3.27}$$

$$= \partial_r \mathbb{N} \dot{r}_p + \partial_s \mathbb{N} \dot{s}_p + \partial_F \mathbb{N} \dot{F}_p + \partial_G \mathbb{N} \dot{G}_p.$$

Using (3.3.17), (3.3.20), (3.3.23), (3.3.24), and the equations of balance (3.3.19) and (3.3.22), we can replace $\dot{r}_p$, $\dot{s}_p$, $\dot{F}_p$, and $\dot{G}_p$ by s_p, $\mathbb{F}(\mathbb{X})/m$, G_p, and $(\mathbb{M}(\mathbb{X}) - \mathbb{S}(\mathbb{X}))E_R^{-1}$, respectively. When the resulting expressions for the derivatives of χ_p and η_p are placed into (3.2.2) and (3.2.3), we obtain the system

$$\partial_r \mathbb{X} s_p + \partial_s \mathbb{X} \mathbb{F}(\mathbb{X})/m + \partial_F \mathbb{X} G_p + \partial_G \mathbb{X}(\mathbb{M}(\mathbb{X}) - \mathbb{S}(\mathbb{X}))E_R^{-1} = \mathbb{N},$$

$$\partial_r \mathbb{N} s_p + \partial_s \mathbb{N} \mathbb{F}(\mathbb{X})/m + \partial_F \mathbb{N} G_p + \partial_G \mathbb{N}(\mathbb{M}(\mathbb{X}) - \mathbb{S}(\mathbb{X}))E_R^{-1}$$

$$= \frac{1}{\rho_R} \operatorname{div} \mathfrak{h}_R(\nabla \mathbb{X}) + b(\mathbb{X}), \qquad \mathbb{X} \in \mathscr{B}, \tag{3.3.28}$$

$$\mathfrak{h}_R(\nabla \mathbb{X})N = t(\mathbb{X}), \qquad \mathbb{X} \in \partial \mathscr{B}.$$

The functions $r_p(t)$, $s_p(t)$, $F_p(t)$, and $G_p(t)$ in this system must satisfy the equations of balance of the theory of pseudo-rigid bodies. However, since we have eliminated all derivatives with respect to t, we may set $t = 0$ and use the arbitrary initial conditions (3.3.18). Thus, the subscript p may be omitted everywhere. We conclude that (3.3.28), together with the restrictions (3.3.16), defines a system of functional differential equations for the determination[7] of the coarsifier $(\mathbb{X}, \mathbb{N})$.

(iv) Symmetry. Uniqueness

The group of rigid-body motions plays a distinguished role in the theory of nonlinear elasticity. It characterizes our perception of the relation between different observers, and so it is built directly into the theory through the statement of the principle of material frame indifference. Material frame indifference requires that all observers use the same response function $\mathfrak{h}_R$. If we denote a general superposed rigid-body motion of the elastic state (χ, η) by

$$\chi^* = z(t) + Q(t)(\chi - x_0), \qquad \eta^* = Q(t)\eta, \tag{3.4.1}$$

$Q(t)$ being an orthogonal tensor and $z(t)$ being a place, then this principle is conventionally formulated as the invariance condition

---

[7] MUNCASTER [1984b] has shown that the representation (3.3.15), the equations of balance (3.3.19) and (3.3.22) for r_p and F_p, and the system (3.3.28) and (3.3.16) for $\mathbb{X}$ and $\mathbb{N}$ are also sufficient conditions for the family determined by χ_p to form a subtheory of nonlinear elasticity.

$$\mathfrak{h}_R(\nabla\chi^*) = Q\mathfrak{h}_R(\nabla\chi). \qquad (3.4.2)$$

Most physical theories are required to satisfy a relation of this type. Indeed, in our direct approach to the theory of pseudo-rigid bodies we see a parallel of this condition in (2.7.11).

Material frame indifference is traditionally imposed *a priori*. The ideas surrounding the notion of consistency *question this tradition*. In the process of examining consistency with nonlinear elasticity, we have, in essence, *derived* a theory of pseudo-rigid bodies. Its response function $\mathfrak{S}_R$ is determined completely by the three-dimensional theory through (3.3.24), and we no longer have the freedom to impose on the derived theory the principle of material frame indifference. Either that principle is true by virtue of (3.4.2), or it is false. We have what appears to be a matter for proof or disproof rather than independent hypothesis. As we shall see shortly, this conjecture is *almost* true.

In order to see how the effects of superposed rigid-body motions filter from nonlinear elasticity down to our derived theory of pseudo-rigid bodies, we must work not in terms of invariance of the stress tensor as given by (3.4.2) but rather in terms of the invariance of the full boundary-vaue problem (3.2.2). Indeed, the framework underlying our approach to consistency is based on properties of solutions of the system (3.2.2). If we can say something about how solutions (i.e., not just the stress) transform under a change of observer, we can attempt to transfer such properties to the solutions that make up our subtheory. Thus, we must first decide in what sense the boundary-value problem (3.2.2) is invariant.

Unfortunately, it is difficult to work with the full group of rigid-body motions and still maintain some simplicity in the calculations. As an alternative, we restrict our attention henceforth in this section to the subgroup of *rigid-body displacements*. That is, Q and z in (3.4.1) are assumed to be independent of time. As a result, we shall not be able to draw conclusions directly about the validity of the principle of material frame indifference for our subtheory. Instead, we consider the restricted version of that principle appropriate to this subgroup, henceforth called the *restricted principle of material frame indifference*. The conclusions, nevertheless, are far reaching and ultimately should serve as a prototype for an analysis based on the full group of rigid-body motions.

Invariance of (3.2.2) means that if χ is a solution, then χ^* is a solution for each choice of Q and z. In order to state this precisely, we introduce first associated transforms of the body force b and surface traction t. These follow the prescription given in (2.6.19) and are given by

$$b^*(x) = Qb(Q^T(x - z) + x_0),$$
$$t^*(x) = Qt(Q^T(x - z) + x_0). \qquad (3.4.3)$$

As an immediate consequence we have the following lemma, which describes precisely the sense in which the group of rigid-body displacements is an *invariance group* for nonlinear elasticity.

Lemma. *If* (χ, η) *is a solution of the boundary-value problem* (3.2.2) *with body force* b *and surface traction* t, *then for any orthogonal tensor* Q *and any place* z *the pair* (χ^*, η^*) *is a solution of this same problem with body force* b^* *and surface traction* t^*.

Proof. By (3.4.1) and (3.4.2) we see that

$$\rho_R \ddot{\chi}^* - \operatorname{div} \mathfrak{h}_R(\nabla \chi^*) - \rho_R b^*(\chi^*)$$
$$= \rho_R Q\dot{\eta} - Q \operatorname{div} \mathfrak{h}_R(\nabla \chi) - \rho_R Qb(Q^T(\chi^* - z) + x_0)$$
$$= Q(\rho_R \ddot{\chi} - \operatorname{div} \mathfrak{h}_R(\nabla \chi) - \rho_R b(\chi)), \qquad (3.4.4)$$

and so $(3.2.2)_1$ is satisfied by χ^* and b^* if it is satisfied by χ and b. The invariance of $(3.2.2)_2$ is verified in the same way. Moreover, (3.4.1) shows directly that $\eta^* = \dot{\chi}^*$ whenever $\eta = \dot{\chi}$.

Since r and F are defined as certain moments of χ, we can calculate directly the effect on them of a superposed rigid-body displacement. By (3.3.3) and (3.3.17) we find that the appropriate transformations are

$$r^* = z - x_0 + Qr, \qquad s^* = Qs, \qquad F^* = QF, \qquad G^* = QG. \quad (3.4.5)$$

In a similar way, we can consider changes in the external force f and the external referential force-moment M_R. These are summarized most conveniently in terms of transformation properties of the associated functionals $\mathbb{F}$ and $\mathbb{M}$. Since it is important to take explicit account now of the body force b and the field of tractions t, let us write $\mathbb{F}(\chi, b, t)$ in place of $\mathbb{F}(\chi)$ and $\mathbb{M}(\chi, b, t)$ in place of $\mathbb{M}(\chi)$. Then (3.3.21) shows that

$$\mathbb{F}(\chi^*, b^*, t^*) = \int_{\partial \mathscr{B}} Qt(Q^T(\chi^* - z) + x_0) + \int_{\mathscr{B}} \rho_R Qb(Q^T(\chi^* - z) + x_0)$$
$$= Q\mathbb{F}(\chi, b, t). \qquad (3.4.6)$$

Similarly, from (3.3.25) we find that

$$\mathbb{M}(\chi^*, b^*, t^*) = Q\mathbb{M}(\chi, b, t). \qquad (3.4.7)$$

Finally, (3.3.26) and (3.4.2) imply the following form of the restricted principle of material frame indifference for the average stress:

$$\mathbb{S}(\chi^*) = Q\mathbb{S}(\chi). \qquad (3.4.8)$$

For the derived theory we obtained for pseudo-rigid bodies, material frame indifference is concerned with transformation properties of the response function $\mathfrak{S}_R$ defined by (3.3.24). More generally, the invariance of the governing differential equations for this subtheory is also related to rules of transformation for the functions $\mathfrak{f}$ and $\mathfrak{M}$ given by (3.3.20) and (3.3.23). While (3.4.6)–(3.4.8) go part way to giving us these, it is clear that we must first consider how $\mathbb{X}$ and $\mathbb{N}$ change owing to a superposed rigid-body displacement.

Since we do not have formulas for $\mathbb{X}$ and $\mathbb{N}$ in terms of other quantities whose laws of transformation are known, we must make direct appeal to the physical problem. The values of $\mathbb{X}$ are places and the values of $\mathbb{N}$ are vectors, so these values should transform according to (3.4.1). At the same time the arguments of $\mathbb{X}$ and $\mathbb{N}$ transform under a rigid-body displacement according to (3.4.3) and (3.4.5). Combining these effects we define $\mathbb{X}^*$ and $\mathbb{N}^*$ by the rules

$$\mathbb{X}^*(r^*, s^*, F^*, G^*, b^*, t^*) = Q(\mathbb{X}(r, s, F, G, b, t) - x_0) + z,$$
$$\mathbb{N}^*(r^*, s^*, F^*, G^*, b^*, t^*) = Q\mathbb{N}(r, s, F, G, b, t).$$
(3.4.9)

Here, as with $\mathbb{F}$ and $\mathbb{M}$, we have included explicit dependence of $\mathbb{X}$ and $\mathbb{N}$ on the body force b and the surface traction t. Finally, if we define $\mathfrak{S}_R^*$, $\mathfrak{f}^*$, and $\mathfrak{M}^*$ in terms of $\mathbb{X}^*$ according to the prescriptions, (3.3.24), (3.3.20), and (3.3.23), respectively, then (3.4.9) and (3.4.6)–(3.4.8) give us

$$\mathfrak{S}_R^*(r^*, s^*, F^*, G^*, b^*, t^*) = Q\mathfrak{S}_R(r, s, F, G, b, t),$$
$$\mathfrak{f}^*(r^*, s^*, F^*, G^*, b^*, t^*) = Q\mathfrak{f}(r, s, F, G, b, t),$$
(3.4.10)
$$\mathfrak{M}^*(r^*, s^*, F^*, G^*, b^*, t^*) = Q\mathfrak{M}(r, s, F, G, b, t).$$

These are the transformation rules for the internal referential force-moment, external force, and external referential force-moment in a rigid-body displacement. They state nothing more than the natural fact that all three quantities transform as vector-valued functions of the condition of a pseudo-rigid body. Indeed, considerations of this type are exactly what led us originally to the transformation (2.6.9).

But what of restricted material frame indifference? Within the context of nonlinear elasticity, we can state a parallel of $(2.7.8)_1$, namely,

$$\mathfrak{h}_R^*(F^*) = Q\mathfrak{h}_R(F),$$
(3.4.11)

and view this as a definition of the response function $\mathfrak{h}_R^*$ to be used by a new observer. It simply states that the first Piola–Kirchhoff stress transforms as a vector, that is, is a frame-indifferent quantity. However, the principle of material frame indifference requires that the elastic material be indistinguishable from frame to frame. Thus, two different observers must use the same response function:

$$\mathfrak{h}_R^* = \mathfrak{h}_R.$$
(3.4.12)

By analogy, within the generalized setting here in which the response function $\mathfrak{S}_R$ depends not only on r, s, F, and G, but also on b and t, material frame indifference would require that

$$\mathfrak{S}_R^* = \mathfrak{S}_R. \tag{3.4.13}$$

We have seen nothing yet that supports this conclusion. Indeed, all the rules of transformation deduced so far follow entirely from the definitions of the theory. We have not yet used the central conclusion of our study of consistency, namely, that $\mathbb{X}$ and $\mathbb{N}$ must be solutions of the equations (3.3.16) and (3.3.28).

The problem for $\mathbb{X}$ and $\mathbb{N}$ enters the analysis of restricted material frame indifference in the following way. We have seen already that, by virtue of (3.4.2), the equations of nonlinear elasticity are invariant; that is, a rigid-body displacement superimposed on any solution produces another solution. As the following result shows, the problem for $\mathbb{X}$ and $\mathbb{N}$ inherits this invariance.

Theorem. If $(\mathbb{X}, \mathbb{N})$ is a solution of (3.3.16) and (3.3.28), then for any orthogonal tensor Q and any place z, $(\mathbb{X}^*, \mathbb{N}^*)$ is also a solution.

Proof. Consider first $(3.3.16)_1$. By hypothesis this is satisfied by $\mathbb{X}$. Thus, $(3.4.9)_1$ shows that

$$\frac{1}{m}\int_{\mathscr{B}} \rho_R(\mathbb{X}^* - x_0) = \frac{1}{m}\int_{\mathscr{B}} \rho_R(Q(\mathbb{X} - x_0) + z - x_0)$$

$$= \frac{1}{m}Q\int_{\mathscr{B}} \rho_R(\mathbb{X} - x_0) + \frac{1}{m}(z - x_0)\int_{\mathscr{B}} \rho_R$$

$$= Qr + z - x_0$$

$$= r^*, \tag{3.4.14}$$

where the arguments of $\mathbb{X}^*$ are the starred variables while those of $\mathbb{X}$ are unstarred. But since $r^*, \ldots, t^*$ are arbitrary, we see that $\mathbb{X}^*$ also satisfies $(3.3.16)_1$. The invariance of $(3.3.16)_{2,3,4}$ is proved in the same way.

For the boundary condition $(3.3.28)_3$, we note from $(3.4.3)_2$, (3.4.2), and $(3.4.9)_1$ that

$$\mathfrak{h}_R(\nabla\mathbb{X}^*)N - t^*(\mathbb{X}^*) = Q\mathfrak{h}_R(\nabla\mathbb{X})N - Qt(Q^T(\mathbb{X}^* - z) + x_0)$$

$$= Q(\mathfrak{h}_R(\nabla\mathbb{X})N - t(\mathbb{X})). \tag{3.4.15}$$

Since $\mathbb{X}$ satisfies the boundary condition, the right-hand side of (3.4.15) vanishes, and so $\mathbb{X}^*$ satisfies it also. By a similar calculation we have the identity

$$\frac{1}{\rho_R}\operatorname{div}\mathfrak{h}_R(\nabla\mathbb{X}^*) + b^*(\mathbb{X}^*) = Q\left(\frac{1}{\rho_R}\operatorname{div}\mathfrak{h}_R(\nabla\mathbb{X}) + b(\mathbb{X})\right). \tag{3.4.16}$$

Finally, consider the left-hand sides of $(3.3.28)_{1,2}$. By direct calculation we see that

$$\partial_{r^*} \mathbb{X}^* s^* = \partial_r \mathbb{X}^* Q^{\mathsf{T}} s^*$$
$$= \partial_r (Q(\mathbb{X} - x_0) + z) Q^{\mathsf{T}} Q s$$
$$= Q \partial_r \mathbb{X} s. \tag{3.4.17}$$

Also, by (3.4.6) we see that

$$\partial_{s^*} \mathbb{X}^* \mathbb{F}(\mathbb{X}^*, b^*, t^*) = \partial_s \mathbb{X}^* Q^{\mathsf{T}} \mathbb{F}(\mathbb{X}^*, b^*, t^*)$$
$$= \partial_s (Q(\mathbb{X} - x_0) + z) Q^{\mathsf{T}} Q \mathbb{F}(\mathbb{X}, b, t)$$
$$= Q \partial_s \mathbb{X} \mathbb{F}(\mathbb{X}, b, t). \tag{3.4.18}$$

Similar calculations apply to each of the terms on the left-hand sides of $(3.3.28)_{1,2}$. If we define the operator D by

$$D(f, K)[h] = \partial_r hs + \partial_s hf + \partial_F hG + \partial_G hK, \tag{3.4.19}$$

we can collect together the foregoing results as the transformations

$$D(\mathbb{F}(\mathbb{X}^*, b^*, t^*)/m, (\mathbb{M}(\mathbb{X}^*, b^*, t^*) - \mathbb{S}(\mathbb{X}^*)) E_R^{-1})[\mathbb{X}^*]$$
$$= Q D(\mathbb{F}(\mathbb{X}, b, t)/m, (\mathbb{M}(\mathbb{X}, b, t) - \mathbb{S}(\mathbb{X})) E_R^{-1})[\mathbb{X}],$$
$$D(\mathbb{F}(\mathbb{X}^*, b^*, t^*)/m, (\mathbb{M}(\mathbb{X}^*, b^*, t^*) - \mathbb{S}(\mathbb{X}^*)) E_R^{-1})[\mathbb{N}^*] \tag{3.4.20}$$
$$= Q D(\mathbb{F}(\mathbb{X}, b, t)/m, (\mathbb{M}(\mathbb{X}, b, t) - \mathbb{S}(\mathbb{X})) E_R^{-1})[\mathbb{N}].$$

From (3.4.16), (3.4.20), and $(3.4.9)_2$, we conclude that $(\mathbb{X}^*, \mathbb{N}^*)$ satisfies $(3.3.28)_{1,2}$ whenever $(\mathbb{X}, \mathbb{N})$ is a solution.

This theorem presents us with a dilemma. It shows that if we find one solution for the coarsifier $(\mathbb{X}, \mathbb{N})$, then we have actually found a whole family, that is, one solution for each Q and z. However, to each $(\mathbb{X}, \mathbb{N})$ there corresponds a derived theory of pseudo-rigid bodies. We therefore obtain multiple theories, all representing coarse descriptions of nonlinear elasticity with the same set of coarse descriptors. In the light of the general question of consistency and the geometric picture we have presented in Section 3(i), this is an unexpected conclusion.

The problem of multiple subtheories vanishes, of course, if the solution of (3.3.16) and (3.3.28) is unique:

$$\mathbb{X}^* = \mathbb{X}, \qquad \mathbb{N}^* = \mathbb{N}. \tag{3.4.21}$$

Moreover, this uniqueness carries immediate consequences for the matter of restricted material frame indifference. Indeed, since $\mathfrak{S}_R = \mathbb{S}(\mathbb{X})$ and $\mathfrak{S}_R^* = \mathbb{S}(\mathbb{X}^*)$, we see directly from (3.4.21) that (3.4.13) holds. In the same way we conclude that

$$\mathfrak{f}^* = \mathfrak{f}, \qquad \mathfrak{M}^* = \mathfrak{M}. \tag{3.4.22}$$

In summary, we have established the following theorem:

Theorem. *If the solution* $(\mathbb{X}, \mathbb{N})$ *of* (3.3.16) *and* (3.3.28) *is unique, then the response function* $\mathfrak{S}_R$ *for the referential internal force-moment satisfies the restricted principle of material frame indifference. Moreover, with respect to the group of rigid-body displacements, the external force and force-moment functions* $\mathfrak{f}$ *and* $\mathfrak{M}$ *are independent of observer, and the associated theory of pseudo-rigid bodies is invariant.*

In view of the types of conclusions we would like to obtain from a study of consistency, this theorem provides strong support for the conjecture that $\mathbb{X}$ and $\mathbb{N}$ are indeed unique. We discuss this conjecture at some length shortly, but first let us note one of its implications. Denote the explicit dependence of b^* and t^* on the rigid-body displacement (3.4.1) by writing

$$b^*(x) = b_{[Q,z]}(x) \equiv Qb(Q^T(x - z) + x_0),$$
$$t^*(x) = t_{[Q,z]}(x) \equiv Qt(Q^T(x - z) + x_0). \tag{3.4.23}$$

Then the requirement of uniqueness (3.4.21), in view of the transformations (3.4.5), (3.4.9), and (3.4.23), becomes

$$\mathbb{X}(Qr + z - x_0, Qs, QF, QG, b_{[Q,z]}, t_{[Q,z]})$$
$$= Q(\mathbb{X}(r, s, F, G, b, t) - x_0) + z,$$
$$\mathbb{N}(Qr + z - x_0, Qs, QF, QG, b_{[Q,z]}, t_{[Q,z]})$$
$$= Q\mathbb{N}(r, s, F, G, b, t). \tag{3.4.24}$$

These must be satisfied for all places z, all orthogonal tensors Q, and of course all values of the arguments r, s, F, G, b, and t. We see in (3.4.24) invariance conditions analogous to those given in (2.7.11) for the response functions $\mathfrak{S}_R$ and σ. Representation theorems led us there to reduced constitutive relations, such as (2.7.13) in which the functions $\mathfrak{S}_u$ and σ^u were arbitrary, and so it is natural here to seek similar reduced functions $\hat{\mathbb{X}}$ and $\hat{\mathbb{N}}$ in terms of which (3.4.24) is satisfied trivially. This can be done by introducing the polar decomposition (2.2.9) for F in which R is orthogonal and U is symmetric and positive-definite. In addition, we have

$$G = \dot{F} = R\dot{U} + \dot{R}U = R(K + \Omega U), \tag{3.4.25}$$

where

$$\Omega = R^T\dot{R}, \qquad K = \dot{U}. \tag{3.4.26}$$

Setting $Q = R^T$ and $z = x_0 - R^Tr$, we obtain from (3.4.24) the representation

$$\mathbb{X}(r, s, F, G, b, t) = R(\hat{\mathbb{X}}(R^Ts, U, K + \Omega U, b_{[R^T, x_0 - R^Tr]}, t_{[R^T, x_0 - R^Tr]}) - x_0)$$
$$+ r + x_0,$$
$$\mathbb{N}(r, s, F, G, b, t) = R\hat{\mathbb{N}}(R^Ts, U, K + \Omega U, b_{[R^T, x_0 - R^Tr]}, t_{[R^T, x_0 - R^Tr]}), \tag{3.4.27}$$

where

$$\hat{\mathbb{X}}(s, U, G, b, t) \equiv \mathbb{X}(0, s, U, G, b, t),$$
$$\hat{\mathbb{N}}(s, U, G, b, t) \equiv \mathbb{N}(0, s, U, G, b, t). \qquad (3.4.28)$$

Note that $\hat{\mathbb{X}}$ and $\hat{\mathbb{N}}$ depend only on the vector s, the symmetric tensor U, the tensor G, the body force b and surface traction t, and of course the suppressed reference place X. Thus, (3.4.27) provides representations of $\mathbb{X}$ and $\mathbb{N}$ in terms of reduced functions $\hat{\mathbb{X}}$ and $\hat{\mathbb{N}}$ depending on fewer variables, and in which the displacement r and the orthogonal transformation R appear explicitly. Moreover, it can easily be shown that (3.4.24) is satisfied trivially for any $\mathbb{X}$ and $\mathbb{N}$ of the form (3.4.27), and so we can henceforth regard $\hat{\mathbb{X}}$ and $\hat{\mathbb{N}}$ as the unknown functions to be determined. We call $(\hat{\mathbb{X}}, \hat{\mathbb{N}})$ the *reduced coarsifier* for the subtheory.

The conjecture of uniqueness raised earlier can be appraised most easily by considering instead the implications of nonuniqueness. If there are multiple solutions $(\mathbb{X}, \mathbb{N})$ of the problem for the coarsifier, then for each one we obtain a theory of pseudo-rigid bodies. From the perspective of the question of consistency that we are exploring here, this might appear rather unsatisfactory. However, there is another way to view the matter. We have founded the analytic study of consistency on the search for a subtheory of nonlinear elasticity parameterized by initial values from the theory of pseudo-rigid bodies. Much of our subsequent development was based, though only implicitly, on the assumption that the closure property (3.2.19) would be enough to single out exactly one coarse theory for pseudo-rigid bodies. If solutions of (3.3.16) and (3.3.28) are unique, this implicit assumption is correct. By the same reasoning, nonuniqueness implies that (3.2.19) is not sufficient, and so additional hypotheses are required. These new hypotheses would serve as *selection rules* for choosing from the multiplicity of coarsifiers $(\mathbb{X}, \mathbb{N})$ one of particular interest for the physical problem. But what form should these additional hypotheses take? We select the possibility suggested by our desire that the constitutive relation of the pseudo-rigid body be materially frame indifferent. By the analysis above, this essentially reduces to imposing (3.4.24), or its equivalent (3.4.21). Note that these no longer assert uniqueness. We are assuming, rather, that *among all possible solutions* $(\mathbb{X}, \mathbb{N})$ *of* (3.3.16) *and* (3.3.28), *we shall select that one which reproduces itself under the action of the group of rigid-body displacements.* We can easily rephrase this assumption in terms of the original subtheory $\mathscr{F}$ given by (3.2.7). If we extend the parameter p to include the body force b and surface traction t, and if we define its transform p^* under a rigid-body displacement according to (3.4.3) and (3.4.5), then we require that *for all places z and all orthogonal transformations Q,*

$$\chi_{p^*}(t) = Q(\chi_p(t) - x_0) + z = \chi_p^*(t). \qquad (3.4.29)$$

With this additional hypothesis (3.4.24) applies, regardless of whether

$\mathbb{X}$ and $\mathbb{N}$ are unique. Of course, uniqueness would guarantee (3.4.29) automatically.

The problem of nonuniqueness is not fully resolved by (3.4.29), and, in fact, we have implicitly introduced a further complication. There is no guarantee that only one coarsifier satisfies this new hypothesis. That is, among all solutions $(\mathbb{X}, \mathbb{N})$ that define a subtheory, *several* may also satisfy (3.4.29). This suggests the need for still further selection rules, based perhaps on other considerations of symmetry from the three-dimensional problem. More important, however, is the new question: In the light of nonuniqueness, *must* there be a solution of (3.3.16) and (3.3.28) with the special property (3.4.24)? More generally, does an invariant problem always have an invariant solution? This represents one of the challenging open problems[8] of the theory of consistency.

(v) Approximation of the Coarsifier

The linkages we have developed between nonlinear elasticity and the theory of pseudo-rigid bodies are provided by the coarsifier $(\mathbb{X}, \mathbb{N})$. Principally, there are three. The representation

$$\chi_p(t) = \mathbb{X}(r(t), \dot{r}(t), F(t), \dot{F}(t)) \tag{3.5.1}$$

gives a direct mapping of solution curves for pseudo-rigid bodies into associated solution curves for an elastic body. The expressions (3.3.20) and (3.3.23) provide us with the external force and external force-moment in terms of the tractions and body forces from the three-dimensional theory. Moreover, with (3.3.24) we can compute the response function $\mathfrak{S}_R$ for pseudo-rigid bodies directly from its parent $\mathfrak{h}_R$ in nonlinear elasticity. The value of these three linkages is contingent, however, on our knowing $\mathbb{X}$ and $\mathbb{N}$. And this, in turn, is contingent on our ability to solve the functional differential system (3.3.16) and (3.3.28).

Apart from our conclusions in Section 3(iv), the analytic questions of

---

[8] In this regard, it is useful to note some facts from the mathematical theory of invariant manifolds. Certain types of invariant manifolds, stable manifolds in particular, are unique (though this is a consequence more of definition than analysis; cf. HALE [1969, §IV.3]). However, for the case of center manifolds there are a number of simple explicit examples that have been put forward to demonstrate nonuniqueness. At the same time, in all the examples we have been able to find, the associated differential equations are invariant under a group, and among the several manifolds that demonstrate nonuniqueness there is one, and only one, with the special symmetry that parallels our requirement (3.4.24). Of course, the explanation of a number of counterexamples does not constitute a proof, but these examples suggest interesting questions for the analyst: For a differential equation that is invariant under a group, is there an invariant manifold that is itself invariant under this group? In regard to nonuniqueness we might ask: If a differential equation exhibits several invariant manifolds, are these manifolds always related to one another by some invariance group for the problem?

existence and uniqueness for this system are largely open. We consider instead a simpler problem, that of generating approximations of $\mathbb{X}$ and $\mathbb{N}$. This might appear to be an unusual point at which to begin, but in fact it is quite natural. In physical theories of even modest complexity, a well-founded scheme of approximation can provide the analyst with important information in regard to existence and uniqueness. This may vary from methods of proof, to choices of function spaces, to restrictions on parameters, or to any other hypotheses that the physics of the problem deems reasonable. A Galerkin approximation might serve as motivation for an analysis of weak solutions, suggesting a space of appropriate functions through the specification of a basis. A Taylor series approximation might play a first step in a proof of local existence, establishing a meaning of "local" in terms of restrictions on external forces, constitutive relations, or solutions themselves. At the same time approximations, when used in place of $\mathbb{X}$ and $\mathbb{N}$ in (3.3.20), (3.3.23), and (3.3.24), would lead to a fully *explicit* theory of pseudo-rigid bodies, one at least approximately consistent with nonlinear elasticity and yet with great practical utility.

Our objective here is to generate Galerkin approximations of $\mathbb{X}$ and $\mathbb{N}$. This type of approximation is naturally suited to the study of consistency because it complements our choices (3.3.3) of the descriptors r and F as averages of quantities from the three-dimensional theory. The Galerkin method essentially considers similar averages of the system governing $\mathbb{X}$ and $\mathbb{N}$. More precisely, $\mathbb{X}$ and $\mathbb{N}$ are approximated by finite linear combinations of functions in a given basis, with coefficients that are weighted averages over the body, and then these coefficients are determined by corresponding averages of the equations (3.3.16) and (3.3.28). We can obtain a preliminary idea of the form of this approximation by assuming that $\mathbb{X}$ and $\mathbb{N}$ are affine functions of X. Then, generalizing (3.3.6), we can show from (3.3.16) that

$$\mathbb{X} = x_0 + r + F(X - \varphi_{\mathrm{R}}),$$
$$\mathbb{N} = s + G(X - \varphi_{\mathrm{R}}). \tag{3.5.2}$$

In this case, (3.3.24) and (3.3.26) simplify to

$$\Sigma_{\mathrm{R}} = \int_{\mathscr{B}} \mathfrak{h}_{\mathrm{R}}(F) = \mathfrak{S}_{\mathrm{R}}(F). \tag{3.5.3}$$

Therefore, if V denotes the volume of the body $\mathscr{B}$ in the reference placement, then $(2.3.8)_2$ and the remarks following (3.2.2) show that $V^{-1}\Sigma$ is the average Cauchy stress over the body. If the elastic material is homogeneous, then (3.5.3) reduces further to

$$\Sigma_{\mathrm{R}} = V\mathfrak{h}_{\mathrm{R}}(F), \tag{3.5.4}$$

and this provides a simple formula for the constitutive relation (2.7.2) for an elastic pseudo-rigid material in terms of the response function for the three-dimensional elastic body. At least for the affine form (3.5.2), the response function $\mathfrak{S}_R$, which generally depends on r, s, F, and G, reduces to a function of F alone. These affine representations also make explicit the expressions (3.3.20) and (3.3.23) for the resultant force f and resultant referential force-moment M_R. The results are (2.4.23) and (2.4.24) recorded previously in Chapter 2.

All these specializations are contingent, of course, on the affine assumptions (3.5.2). Consistency requires that in place of such an assumption we appeal directly to the system (3.3.16) and (3.3.28). Our aim here is to show that, at least for a certain class of Galerkin approximations, the expressions (3.5.2) always arise naturally to a first approximation. In this sense, the preceding specializations provide a natural means of obtaining from our analysis a theory of pseudo-rigid bodies with practical utility for applications.

Using the operator D defined by (3.4.19), we can write (3.3.16) and (3.3.28) in the compact form

$$D(\mathbb{F}(\mathbb{X})/m, (\mathbb{M}(\mathbb{X}) - \mathbb{S}(\mathbb{X}))E_R^{-1})[\mathbb{X}] = \mathbb{N},$$

$$D(\mathbb{F}/m, (\mathbb{M}(\mathbb{X}) - \mathbb{S}(\mathbb{X}))E_R^{-1})[\mathbb{N}] = \frac{1}{\rho_R} \operatorname{div} \mathfrak{h}_R(\nabla \mathbb{X}) + b(\mathbb{X}), \qquad X \in \mathscr{B},$$

$$\mathfrak{h}_R(\nabla \mathbb{X})N = t(\mathbb{X}), \qquad X \in \partial\mathscr{B}, \tag{3.5.5}$$

$$r = \frac{1}{m} \int_{\mathscr{B}} \rho_R(\mathbb{X} - x_0), \qquad F = \int_{\mathscr{B}} \rho_R(\mathbb{X} - x_0) \otimes (X - \varphi_R)E_R^{-1},$$

$$s = \frac{1}{m} \int_{\mathscr{B}} \rho_R \mathbb{N}, \qquad G = \int_{\mathscr{B}} \rho_R \mathbb{N} \otimes (X - \varphi_R)E_R^{-1}.$$

The first two of these have the appearance of a set of differential equations, but since $\mathbb{F}$, $\mathbb{M}$, and $\mathbb{S}$ are the functions of χ given by (3.3.21), (3.3.25), and (3.3.26), implicitly they also contain certain integral operators. Fortunately, this added complication can be removed, leading to a slightly simpler problem.

Theorem. *If* $f: \mathscr{P} \to \mathscr{V}$, $K: \mathscr{P} \to \mathscr{V} \otimes \mathscr{V}$, $\mathbb{X}: \mathscr{B} \times \mathscr{P} \to \mathscr{E}$, *and* $\mathbb{N}: \mathscr{B} \times \mathscr{P} \to \mathscr{V}$ *are any smooth functions that satisfy the system of equations*

$$D(f, K)[\mathbb{X}] = \mathbb{N},$$

$$D(f, K)[\mathbb{N}] = \frac{1}{\rho_R} \operatorname{div} \mathfrak{h}_R(\nabla \mathbb{X}) + b(\mathbb{X}), \qquad X \in \mathscr{B},$$

$$\mathfrak{h}_R(\nabla \mathbb{X})N = t(\mathbb{X}), \qquad X \in \partial\mathscr{B}, \tag{3.5.6}$$

$$r = \frac{1}{m} \int_{\mathscr{B}} \rho_R(\mathbb{X} - x_0), \qquad F = \int_{\mathscr{B}} \rho_R(\mathbb{X} - x_0) \otimes (X - \varphi_R)E_R^{-1},$$

then

$$f = \frac{1}{m} \mathbb{F}(\mathbb{X}), \qquad K = (\mathbb{M}(\mathbb{X}) - \mathbb{S}(\mathbb{X}))E_R^{-1}, \qquad (3.5.7)$$

and $\mathbb{X}$ *and* $\mathbb{N}$ *satisfy* (3.5.5).

Thus, only the first and second of the restrictions $(3.5.5)_{4-7}$ need be considered and we may view $\mathbb{F}$ and $\mathbb{M} - \mathbb{S}$ in $(3.5.5)_{1,2}$ as additional unknowns f and K rather than as functionals of χ.

Proof. By (3.4.19) we note that

$$\int_{\mathscr{B}} \rho_R D(f, K)[h] = D(f, K)\left[\int_{\mathscr{B}} \rho_R h\right],$$
$$\int_{\mathscr{B}} \rho_R D(f, K)[h] \otimes (X - \varphi_R) = D(f, K)\left[\int_{\mathscr{B}} \rho_R h \otimes (X - \varphi_R)\right]. \qquad (3.5.8)$$

Setting $h = \mathbb{X} - x_0$ and using $(3.5.6)_{1,4,5}$, we obtain

$$\int_{\mathscr{B}} \rho_R \mathbb{N} = D(f, K)[mr] = ms,$$
$$\int_{\mathscr{B}} \rho_R \mathbb{N} \otimes (X - \varphi_R) = D(f, K)[FE_R] = GE_R. \qquad (3.5.9)$$

These are precisely the restrictions $(3.5.5)_{6,7}$. Next, let us set $h = \mathbb{N}$ and use $(3.5.6)_{2,4,5}$. By the divergence theorem and the boundary conditions, we find that

$$\mathbb{F}(\mathbb{X}) = D(f, K)[ms]$$
$$= mf,$$
$$\mathbb{M}(\mathbb{X}) - \mathbb{S}(\mathbb{X}) = D(f, K)[GE_R] \qquad (3.5.10)$$
$$= KE_R,$$

and these are (3.5.7). It is now clear that $\mathbb{X}$ and $\mathbb{N}$ satisfy (3.5.5). $\blacksquare$

Consider a weak formulation of the problem posed by $(3.5.6)_{1,2,3}$. If $\mathfrak{h}_R = (\mathfrak{h}_i)$ denotes the decomposition of the stress into traction vectors relative to a fixed basis, then we multiply $(3.5.6)_{1,2}$ by $\rho_R \psi$, with ψ denoting any smooth real-valued function on the closure of $\mathscr{B}$, and integrate over $\mathscr{B}$. By the divergence theorem and the boundary conditions $(3.5.6)_3$, we obtain

$$D(f, K)\left[\int_{\mathscr{B}} \rho_R(\mathbb{X} - x_0)\psi\right] = \int_{\mathscr{B}} \rho_R \mathbb{N}\psi,$$

$$D(f, K)\left[\int_{\mathscr{B}} \rho_R \mathbb{N}\psi\right] = \int_{\partial\mathscr{B}} t(\mathbb{X})\psi + \int_{\mathscr{B}} \rho_R b(\mathbb{X})\psi - \int_{\mathscr{B}} \mathfrak{h}_i(\nabla\mathbb{X})\psi_{,i}. \qquad (3.5.11)$$

For ψ we choose the components of a set of tensor-valued polynomials $\psi^{(n)}$ of X formed as follows. Set

$$\psi^{(0)} = 1, \qquad \psi^{(1)} = X - \varphi_R. \tag{3.5.12}$$

Then form $\psi^{(n)}$ by the Gram–Schmidt orthogonalization procedure, namely, by subtracting from $X \otimes \cdots \otimes X$ (n factors) its projections along $\psi^{(0)}, \psi^{(1)}, \ldots, \psi^{(n-1)}$ so that

$$\int_{\mathscr{B}} \rho_R \psi^{(k)} \otimes \psi^{(l)} = 0 \quad \text{for } k \neq l. \tag{3.5.13}$$

For $k = 0$ and $l = 1$, this orthogonality condition is guaranteed by (3.3.2).

The nth Galerkin approximations of X and N are given by

$$X^{(n)} = x_0 + \sum_{k=0}^{n} A_k^{(n)} \psi^{(k)},$$
$$N^{(n)} = \sum_{k=0}^{n} B_k^{(n)} \psi^{(k)}, \tag{3.5.14}$$

where the tensor coefficients $A_k^{(n)}$ and $B_k^{(n)}$ and the auxiliary variables $f^{(n)}$ and $K^{(n)}$ are those functions of r, s, F, and G for which

$$D(f^{(n)}, K^{(n)}) \left[\int_{\mathscr{B}} \rho_R(X^{(n)} - x_0) \otimes \psi^{(l)} \right]$$
$$= \int_{\mathscr{B}} \rho_R N^{(n)} \otimes \psi^{(l)},$$
$$D(f^{(n)}, K^{(n)}) \left[\int_{\mathscr{B}} \rho_R N^{(n)} \otimes \psi^{(l)} \right] \tag{3.5.15}$$
$$= \int_{\partial\mathscr{B}} t(X^{(n)}) \otimes \psi^{(l)} + \int_{\mathscr{B}} \rho_R b(X^{(n)}) \otimes \psi^{(l)} - \int_{\mathscr{B}} \mathfrak{h}_i(\nabla X^{(n)}) \otimes \psi^{(l)}_{,i},$$
$$r = \frac{1}{m} \int_{\mathscr{B}} \rho_R(X^{(n)} - x_0), \qquad F = \int_{\mathscr{B}} \rho_R(X^{(n)} - x_0) \otimes (X - \varphi_R) E_R^{-1},$$

for $l = 0, 1, \ldots, n$. For this system of equations, we have the following theorem:

Theorem. *The Galerkin approximations of X and N have the form*

$$X^{(n)} = x_0 + r + F(X - \varphi_R) + \sum_{k=2}^{n} A_k^{(n)} \psi^{(k)},$$
$$N^{(n)} = s + G(X - \varphi_R) + \sum_{k=2}^{n} B_k^{(n)} \psi^{(k)},$$
$$mf^{(n)} = \mathbb{F}(X^{(n)}), \tag{3.5.16}$$
$$K^{(n)} E_R = \mathbb{M}(X^{(n)}) - \mathbb{S}(X^{(n)}),$$

where $A_k^{(n)}$ and $B_k^{(n)}$ satisfy (3.5.15) for $l = 2, 3, \ldots, n$.

Proof. By (3.5.14) and the orthogonality relations (3.5.13) we find that

$$\int_{\mathscr{B}} \rho_R(\mathbb{X}^{(n)} - x_0) \otimes \psi^{(l)} = \sum_{k=0}^{n} A_k^{(n)} \int_{\mathscr{B}} \rho_R \psi^{(k)} \otimes \psi^{(l)}$$

$$= A_l^{(n)} \int_{\mathscr{B}} \rho_R \psi^{(l)} \otimes \psi^{(l)},$$

$$\int_{\mathscr{B}} \rho_R \mathbb{N}^{(n)} \otimes \psi^{(l)} = \sum_{k=0}^{n} B_k^{(n)} \int_{\mathscr{B}} \rho_R \psi^{(k)} \otimes \psi^{(l)}$$

$$= B_l^{(n)} \int_{\mathscr{B}} \rho_R \psi^{(l)} \otimes \psi^{(l)}.$$

(3.5.17)

Using (3.3.1), (3.5.12), and (3.5.15)$_1$, we see that these reduce for $l = 0$ and $l = 1$ to

$$A_0^{(n)} = r, \qquad A_1^{(n)} = F,$$

$$B_0^{(n)} = D(f^{(n)}, K^{(n)})[mr]/m = s,$$

$$B_1^{(n)} = D(f^{(n)}, K^{(n)})[FE_R]E_R^{-1} = G.$$

(3.5.18)

This completes the proof of (3.5.16)$_{1,2}$ and shows that (3.5.15) is satisfied for $l = 0$ and $l = 1$. Moreover, for these values of l, (3.5.15)$_2$ reduces to

$$\mathbb{F}(\mathbb{X}^{(n)}) = D(f^{(n)}, K^{(n)})[ms]$$

$$= mf^{(n)},$$

$$\mathbb{M}(\mathbb{X}^{(n)}) - \mathbb{S}(\mathbb{X}^{(n)}) = D(f^{(n)}, K^{(n)})[GE_R]$$

$$= K^{(n)}E_R,$$

(3.5.19)

and these give us (3.5.16)$_{3,4}$.

CHAPTER 4

Explicit Motions of Pseudo-rigid Bodies

The preceding chapters lay the foundations of the theory of pseudo-rigid bodies and result in governing differential equations. Here we examine some special pseudo-rigid motions by seeking solutions to these equations. Special solutions provide one means of appraising the quality of our developments: they permit comparisons not only with more familiar results but also with our intuition and experience concerning particular problems. In order to obtain a variety of *explicit* pseudo-rigid motions, we proceed by the *semi-inverse method*. Namely, we speculate about the general structure or form of motions in a given class, informed by experience from rigid-body mechanics and nonlinear elasticity, and then we seek conditions under which this structure prevails in time.

(i) Rigid Motions

It is natural to begin the study of pseudo-rigid motions by considering rigid motions. Thus, we restrict the deformation tensor F to be a rotation R. Moreover, we assume the center of mass P of $\mathscr{B}_e$ to be non-accelerating, and so the resultant force f vanishes. Following previous convention, we consider only elastic pseudo-rigid bodies, and for convenience here we assume the reference placement of the body to be natural: $\mathfrak{S}_R(I) = 0$. From $(2.7.11)_1$, it follows that $\mathfrak{S}_R(R) \equiv 0$. In view of $(2.2.4)_2$, $(2.3.8)_1$, and the foregoing assumptions, the general equations of motion $(2.3.9)_{3,4}$ reduce to

$$\ddot{R}E_R R^{\mathsf{T}} = M. \tag{4.1.1}$$

We have noted previously that the skew part of (4.1.1) corresponds to the standard vector form (2.5.36) of EULER's equation of motion for rigid bodies. Indeed, if we rewrite (4.1.1) as

$$\mathrm{sk}(\ddot{R}E_R R^{\mathsf{T}}) = \mathrm{sk}\ M, \qquad \mathrm{sym}(\ddot{R}E_R R^{\mathsf{T}}) = \mathrm{sym}\ M, \tag{4.1.2}$$

then $(4.1.2)_1$ is a tensor differential equation for the determination of R

while $(4.1.2)_2$ is an algebraic equation for the symmetric force-moment required to sustain the rigid motion. Restricting our attention further to torque-free motions, we assume henceforth that sk $M = 0$. The governing equations (4.1.2) then reduce to

$$\text{sk}(\ddot{R}E_R R^T) = 0, \qquad M = \text{sym}(\ddot{R}E_R R^T). \qquad (4.1.3)$$

Since $(4.1.3)_1$ is a tensor equation, it provides a nonstandard version of EULER'S equations for a rigid body. In this format the rotation R appears directly, and so the reference placement plays a role not usually inherent in the standard treatment based on the angular velocity vector w. As we shall see this suggests an equally nonstandard introduction of the Euler angles. Of course this nonstandard approach ultimately yields conventional and well-known results, but the new perspective given to these is enlightening.

We begin by recalling some notation and results from multilinear algebra. The standard decomposition of second-order tensors on $\mathscr{V}$ into symmetric and skew parts can be expressed as the direct sum

$$\mathscr{V} \otimes \mathscr{V} = \mathscr{S} \oplus \mathscr{A}, \qquad (4.1.4)$$

where $\mathscr{S} = \text{sym}(\mathscr{V} \otimes \mathscr{V})$ and $\mathscr{A} = \text{sk}(\mathscr{V} \otimes \mathscr{V})$. Let Sk: $\mathscr{V} \to \mathscr{A}$ be the linear isomorphism defined by the identity

$$(\text{Sk } a)v = a \times v \quad \text{for all } v \in \mathscr{V}. \qquad (4.1.5)$$

Thus, Sk a is the skew tensor associated with the vector a. Equivalently, a is the axial vector of the skew tensor Sk a. Thus, ax(Sk a) = a, and therefore (Sk a)a = 0. Let the vectors D_i, the directors *fixed* in the reference placement, be members of an orthonormal basis of $\mathscr{V}$. Then the tensors A_i = Sk D_i form an associated basis on $\mathscr{A}$. It is elementary to show that

$$A_i = \varepsilon_{ikj} D_j \otimes D_k \qquad (4.1.6)$$

and hence that the A_i are orthogonal and have norm $\sqrt{2}$ with respect to the standard inner product[1] on $\mathscr{V} \otimes \mathscr{V}$. We define a basis D_{ij} on $\mathscr{S}$ associated with D_i by the equation

$$D_{ij} = \text{sym } D_i \otimes D_j \qquad (4.1.7)$$

and note that the D_{ij} are also orthogonal on $\mathscr{V} \otimes \mathscr{V}$.

With each fixed skew tensor W we may associate a time-dependent rotation $R(t)$ according to the formula

$$R(t) = \exp tW = \sum_{n=0}^{\infty} \left(\frac{t^n}{n!}\right) W^n. \qquad (4.1.8)$$

The properties of the exponential function exp: $\mathscr{A} \to \mathcal{O}(\mathscr{V})$ are well known.

---

[1] This inner product is $T \cdot S = \text{tr } TS^T$, and the corresponding norm is $\|T\| = (\text{tr } TT^T)^{1/2}$.

In particular, R satisfies the differential equation

$$\dot{R} = WR, \tag{4.1.9}$$

where

$$W = \dot{R}R^T, \tag{4.1.10}$$

W being the angular velocity tensor introduced following (2.2.10). Since this tensor and hence the angular velocity vector $w = \text{ax } W$ are independent of time, R is a rotation about an *axis fixed in space*. We employ such rotations repeatedly in the analysis to follow. More generally, let us set

$$Q_i(\alpha) = \exp \alpha A_i. \tag{4.1.11}$$

Since A_i is skew, each of its powers can easily be expressed in terms of I, A_i, and A_i^2, and then the infinite series for Q_i can be summed explicitly. This produces the simple formula

$$Q_i(\alpha) = I + \sin \alpha \, A_i + (1 - \cos \alpha)A_i^2. \tag{4.1.12}$$

In the analysis to follow, we choose the director basis to lie along the principal axes of the referential Euler tensor E_R. We remind the reader that the D_i are body axes in the reference placement. Since they are fixed, they also serve as a set of spatial axes for the current placement. The current values of the directors are given by $d_i = RD_i$, and they lie along the principal directions of E. The d_i constitute the body axes in the present placement.

CASE 1 (Symmetric bodies). Consider first a completely symmetric body:

$$E = E_R = EI. \tag{4.1.13}$$

We look for solutions of $(4.1.3)_1$ in the form

$$R(t) = \exp \theta(t)A, \qquad W = \dot{\theta}A, \tag{4.1.14}$$

where A is a fixed skew tensor and $\theta(t)$ is a real-valued function of time. Thus, (4.1.14) is the generalization of (4.1.8) to a variable-speed fixed-axis rotation. From (4.1.14) we find that

$$\dot{R} = R(\dot{\theta}A), \qquad \ddot{R} = R(\dot{\theta}^2A^2 + \ddot{\theta}A), \tag{4.1.15}$$

and so $(4.1.3)_1$ reduces to

$$\dot{\theta}^2(A^2E_R - E_RA^2) + \ddot{\theta}(AE_R + E_RA) = 0. \tag{4.1.16}$$

By $(4.1.13)_2$ we conclude that

$$\ddot{\theta}A = 0, \quad \text{and so} \quad \theta(t) = at + b, \qquad a = \text{const.}, \, b = \text{const.} \tag{4.1.17}$$

This is the classical result that *a symmetric body can rotate about any given axis with any given constant angular speed* $\dot{\theta} = a$.

We can actually show for this case that R *must* have a fixed angular velocity; that is, $W = \dot{R}R^T$ must be constant. Indeed,

$$W = \ddot{R}R^T + \dot{R}\dot{R}^T$$

$$= \text{sym}(\ddot{R}R^T) + \dot{R}\dot{R}^T + \text{sk}(\ddot{R}R^T)$$

$$= \tfrac{1}{2}(RR^T)^{\cdot\cdot} + \text{sk}(\ddot{R}R^T). \tag{4.1.18}$$

The first term vanishes because R is orthogonal. The second term vanishes by $(4.1.3)_1$. Thus, W is constant.

CASE 2 (Rotation about a principal axis of E_R). Let

$$E_R = \sum_{i=1}^3 E_i D_{ii}, \tag{4.1.19}$$

where each $E_i > 0$. Consider any one of the rotations

$$R_i = Q_i(\theta_i(t)) = \exp \theta_i(t) A_i, \tag{4.1.20}$$

$i = 1$, 2, or 3, where the fixed axis of rotation is the principal axis D_i. An elementary calculation using (4.1.7), (4.1.9), and (4.1.19) shows that for $i \neq j \neq k$

$$A_i^2 E_R - E_R A_i^2 = 0,$$

$$A_i E_R + E_R A_i = (E_j + E_k)A_i. \tag{4.1.21}$$

Therefore (4.1.16), with A replaced by A_i, reduces once again to (4.1.17). *A (generally) nonsymmetric body can rotate about any principal axis of inertia with any given constant angular speed.* This result too is classical.

CASE 3 (Precession of bodies with an axis of symmetry). Consider next solutions R that are a product of two rotations, each with a fixed axis. This type of solution is common when the body has an axis of symmetry, say along D_1. Thus, E_R will have the form

$$E_R = E_1 D_{11} + E(D_{22} + D_{33}), \tag{4.1.22}$$

where $E_1 > 0$ and $E > 0$. Let s be the axis of rotation D_1, and for convenience we set $S = A_1$. Denote the other axis of rotation by the unit vector p and let $P = \text{Sk } p$. There is no loss in generality in assuming that p lies in the plane of D_1 and D_3. Thus,

$$p = \cos \alpha \, s - \sin \alpha \, D_3, \qquad P = \cos \alpha \, S - \sin \alpha \, A_3, \tag{4.1.23}$$

where α is the angle between p and s. In terms of these two vectors, we look for solutions of $(4.1.3)_1$ of the form

$$R = R_p R_s, \tag{4.1.24}$$

where

$$R_p(t) = \exp p(t)P, \qquad R_s(t) = \exp s(t)S. \qquad (4.1.25)$$

The instantaneous axis of rotation of R is neither a principal axis of E_R nor a fixed direction in space. Indeed, the angular velocity tensor is

$$
\begin{aligned}
W &= \dot{R}R^T \\
&= (\dot{R}_p R_s + R_p \dot{R}_s) R_s^T R_p^T \\
&= (\dot{p}P R_p R_s + \dot{s} R_p S R_s) R_s^T R_p^T \\
&= \dot{p}P + \dot{s}R_p SR_p^T, \qquad (4.1.26)
\end{aligned}
$$

and so the angular velocity vector is

$$w = \dot{p}p + \dot{s}R_p s. \qquad (4.1.27)$$

Solutions of this type represent a spinning body precessing about the axis p with precession rate $\dot{p}$.

Placing (4.1.24) and (4.1.25) into (4.1.3)$_1$ and using results analogous to (4.1.15), we obtain the reduced form

$$\mathrm{sk}\{R_p[\dot{p}^2P^2 + \ddot{p}P + 2(\dot{p}P)(\dot{s}S) + \dot{s}^2S^2 + \ddot{s}S]R_s E_R R_s^T R_p^T\} = 0. \quad (4.1.28)$$

Equations (4.1.22) and (4.1.25)$_2$ show that $R_s E_R R_s^T = E_R$, so R_s and E_R commute. This is simply another way to say that E_R has s as an axis of symmetry. Also $\mathrm{sk}(QAQ^T) = Q(\mathrm{sk}\,A)Q^T$ for any tensors Q and A, so (4.1.28) reduces further to

$$\dot{p}^2\,\mathrm{sk}(P^2E_R) + \ddot{p}\,\mathrm{sk}(PE_R) + 2\dot{p}\dot{s}\,\mathrm{sk}(PSE_R) + \dot{s}^2\,\mathrm{sk}(S^2E_R) + \ddot{s}\,\mathrm{sk}(SE_R) = 0. \tag{4.1.29}$$

By (4.1.6) and (4.1.22) the second last term vanishes and the last term is simply $\ddot{s}ES$. To simplify the rest we use the identity

$$PS = s \otimes p - (s \cdot p)I, \qquad (4.1.30)$$

which can be derived from (4.1.6) and (4.1.23). It is elementary, though tedious, to show now that

$$
\begin{aligned}
\mathrm{sk}(PSE_R) &= -\tfrac{1}{2}E \sin \alpha\, A_2, \\
\mathrm{sk}(P^2E_R) &= -\tfrac{1}{2}(E - E_1) \sin \alpha \cos \alpha\, A_2, \qquad (4.1.31) \\
\mathrm{sk}(PE_R) &= -\tfrac{1}{2}(E + E_1) \sin \alpha\, A_3 + E \cos \alpha\, S.
\end{aligned}
$$

Therefore, (4.1.29) simplifies to

$$\ddot{p}\,\mathrm{sk}(PE_R) + \ddot{s}ES - \tfrac{1}{2}\dot{p}\sin\alpha\,[2E\dot{s} + (E - E_1)\dot{p}\cos\alpha]A_2 = 0. \quad (4.1.32)$$

Finally, we note that the skew tensors $\mathrm{sk}(PE_R)$, S, and A_2 are linearly independent, and so we obtain

$$\ddot{p} = 0, \qquad \ddot{s} = 0, \qquad \dot{s} = \frac{1}{2}\left(\frac{E_1 - E}{E}\right)\dot{p}\cos\alpha. \qquad (4.1.33)$$

These results are classical. They show that *in precession the angular velocities about both axes p and s are constant, and these are related to one another and the angle α between the axes through* (4.1.33)$_3$. The factor $\frac{1}{2}$ in that relation might appear incorrect. However, in the classical approach it is more customary to work with the inertia tensor J_R defined analogous to (2.1.21). By (4.1.22), J_R takes the same form as E_R and its principal values J_1 and J are related to those of E_R as follows:

$$J_1 = 2E, \qquad J = E_1 + E. \tag{4.1.34}$$

With these (4.1.33)$_3$ becomes precisely the conventional result

$$\dot{s} = \frac{(J - J_1)}{J_1} \dot{p} \cos \alpha. \tag{4.1.35}$$

Two additional features of this solution should be noted. By (2.2.21), (2.2.28)$_1$, and the fact that $F = R$, the moment of momentum tensor H in the present placement is given by

$$H = \dot{R} E_R R^T. \tag{4.1.36}$$

Recall that the usual angular momentum is the axial vector corresponding to the skew part of H. Noting that

$$\dot{R} = R_p (\dot{p}P + \dot{s}S) R_s \tag{4.1.37}$$

and recalling that R_s and E_R commute, we find that

$$\text{sk } H = R_p \text{ sk}(\dot{p}PE_R + \dot{s}SE_R) R_p^T$$
$$= R_p(-\tfrac{1}{2}(E + E_1)\dot{p} \sin \alpha \, A_3 + E(\dot{p} \cos \alpha + \dot{s})S) R_p^T, \tag{4.1.38}$$

the latter step being a consequence of (4.1.21)$_2$ and (4.1.31)$_3$. Eliminating $\dot{s}$ by the use of (4.1.33)$_3$, and recalling (4.1.23)$_2$ and (4.1.34), we obtain

$$\text{sk } H = \tfrac{1}{2}(E + E_1)\dot{p}P = \tfrac{1}{2}J\dot{p}P. \tag{4.1.39}$$

The angular momentum in precession is constant and is parallel to the axis of precession p.

Perhaps of more interest here is the expression for the force-moment M required to produce a rigid motion. It can be calculated directly from (4.1.3)$_2$ and (4.1.24). The result is

$$M = -\dot{p}R_p \left\{ E_1 \sin^2 \alpha \, D_{11} + \frac{1}{4E}[E^2 + 3E^2 \sin^2 \alpha \right.$$

$$+ E_1(E_1 + 2E)\cos^2 \alpha]D_{22} + \frac{1}{4E}(E + E_1)^2 \cos^2 \alpha \, D_{33}$$

$$\left. + 2E_2 \sin \alpha \cos \alpha \, D_{13} \right\} R_p^T. \tag{4.1.40}$$

The three diagonal terms represent compressive forces in the directions of the principal axes of the Euler tensor. The term in D_{13} contributes a double force without torque and prevents shearing of the body in the plane of D_1 and D_3.

CASE 4 (General rotations and the Euler angles). Generalizing the analysis of Case 3, we consider now solutions R that are products of three rotations, each with a fixed axis. The Euler tensor may be nonsymmetric as in Case 2. One special choice of the axes of rotation is especially important. In terms of the orthonormal directors D_i we consider a rotation R defined by

$$R = R_1 R_2 R_3,$$
$$R_1(t) = Q_1(\varphi(t)), \qquad R_2(t) = Q_2(\theta(t)), \qquad R_3(t) = Q_1(\psi(t)),$$

(4.1.41)

where $Q_i(\alpha)$ is the rotation through angle α about axis D_i given by (4.1.11). The instantaneous angular velocity tensor for R is

$$W = \dot\varphi A_1 + \dot\theta R_1 A_2 R_1^{\mathrm{T}} + \dot\psi R_1 R_2 A_1 R_2^{\mathrm{T}} R_1^{\mathrm{T}},$$

(4.1.42)

and so the angular velocity vector is

$$w = \dot\varphi D_1 + \dot\theta R_1 D_2 + \dot\psi R_1 R_2 D_1.$$

(4.1.43)

The body axes d_i in the current placement are given by $d_i = RD_i$. The angular velocity vector w has components $w_i = w \cdot d_i$ with respect to these axes. From (4.1.41) and (4.1.43) we find that

$$w_1 = \dot\varphi D_1 \cdot R_2^{\mathrm{T}} D_1 + \dot\psi,$$
$$w_2 = \dot\varphi D_2 \cdot R_3^{\mathrm{T}} R_2^{\mathrm{T}} D_1 + \dot\theta D_2 \cdot R_3^{\mathrm{T}} D_2,$$
$$w_3 = \dot\varphi D_3 \cdot R_3^{\mathrm{T}} R_2^{\mathrm{T}} D_1 + \dot\theta D_3 \cdot R_3^{\mathrm{T}} D_2,$$

(4.1.44)

where we have used the fact that R_1 and R_3 have axis D_1 and R_2 has axis D_2. The inner products above can be calculated directly from (4.1.12), and we arrive at

$$w_1 = \dot\varphi \cos\theta + \dot\psi,$$
$$w_2 = \dot\varphi \sin\psi \sin\theta + \dot\theta \cos\psi,$$
$$w_3 = \dot\varphi \cos\psi \sin\theta - \dot\theta \sin\psi.$$

(4.1.45)

These are the classical body-axis components of angular velocity in terms of the Euler angles φ, θ, and ψ. We conclude that (4.1.41) is a representation of a general rotation in three dimensions, and through it we can analyze directly in terms of rotations any motion of a rigid body, indeed, any rigid motion of the pseudo-rigid bodies being considered here.

Geometrically, the interpretation of the product (4.1.41) in terms of the Euler angles requires a further observation. Let us define

$$\bar{R} = \bar{R}_3 \bar{R}_2 \bar{R}_1, \tag{4.1.46}$$

where

$$\bar{R}_1 = R_1, \qquad \bar{R}_2 = R_1 R_2 R_1^\mathsf{T}, \qquad \bar{R}_3 = \bar{R}_2 R_3 \bar{R}_2^\mathsf{T}. \tag{4.1.47}$$

Using the fact that

$$R_1 R_3 = R_3 R_1, \tag{4.1.48}$$

we find from (4.1.47) that

$$\bar{R}_3 = R_1 R_2 R_3 R_2^\mathsf{T} R_1^\mathsf{T}. \tag{4.1.49}$$

Thus, substituting (4.1.47)$_1$, (4.1.47)$_2$, and (4.1.49) in (4.1.46) we conclude that

$$\bar{R} = R. \tag{4.1.50}$$

This shows that there are two equivalent decompositions of the general rotation R. That given by (4.1.46) and (4.1.47) is the standard definition of the Euler angles.[2] The axis of the first rotation $\bar{R}_1$ is D_1 and the angle of rotation is φ. The axis of the second rotation $\bar{R}_2$ is $R_1 D_2$ and the angle of rotation is θ. Finally, the axis of the third rotation $\bar{R}_3$ is $\bar{R}_2 \bar{R}_1 D_1 = R_1 R_2 D_1$ and the angle of rotation is ψ.

Other choices of the Euler angles can be defined in a similar fashion. For example, in the field of aerospace science it is more conventional to consider a general rotation as a product of rotations about three different axes.[3] Indeed, these are chosen along D_3, D_2, and D_1, respectively, and so

$$R = R_1 R_2 R_3, \qquad R_i(t) = Q_i(\varphi_i(t)), \qquad i = 1, 2, 3. \tag{4.1.51}$$

The angular velocity w and its components w_i relative to body axes are given by the natural analogues of (4.1.43) and (4.1.44), respectively. When the explicit form (4.1.12) of Q_i is used to compute the inner products in (4.1.44), we obtain the associated companion of (4.1.45), namely,

$$w_1 = \dot{\varphi}_1 \cos \varphi_2 \cos \varphi_3 + \dot{\varphi}_2 \sin \varphi_3,$$

$$w_2 = -\dot{\varphi}_1 \cos \varphi_2 \sin \varphi_3 + \dot{\varphi}_2 \cos \varphi_3, \tag{4.1.52}$$

$$w_3 = \dot{\varphi}_1 \sin \varphi_2 + \dot{\varphi}_3.$$

---

[2] For one of the better graphics associated with the definition of the Euler angles, we refer the reader to GOLDSTEIN [1980, p. 146]. The vectors d_1, d_2, and d_3 (D_1, D_2, and D_3) correspond to his axes z', x', and y' (z, x, and y), respectively.

[3] This definition of the Euler angles can be found in GREENWOOD [1965], though he writes $R = R_3 R_2 R_1$ in place of our (4.1.51)$_1$.

(ii) **Roto-deformations**

In general, any pseudo-rigid body exhibits some degree of deformation in the course of its motion. The polar decomposition

$$F = RU \tag{4.2.1}$$

provides a convenient means of dividing the deformation into a rotation part R and a strain part as characterized by the (right) stretch tensor U. The simplest motions exhibiting deformation are those for which $R \equiv I$. These are the *pure stretch* solutions and we examine them in Section 4(iii). Here we consider the simplest nontrivial class of solutions in which there is a direct interplay between rotation and stretching. This is the class of *roto-deformations* in which the center of mass of the body is fixed and rotation occurs only about a fixed axis. The latter assumption implies that

$$R = \exp \theta(t)A, \tag{4.2.2}$$

where A is a fixed skew tensor. Without loss of generality, we choose A to be A_1, and so rotation occurs about the axis defined by the fixed director D_1.

The equation of motion $(2.3.9)_4$ can be written in the form

$$\ddot{F}E_R F^T = M - \Sigma \tag{4.2.3}$$

in terms of the conventions $(2.3.8)_{1,2}$, namely,

$$M = M_R F^T, \qquad \Sigma = \Sigma_R F^T. \tag{4.2.4}$$

By (4.2.1) and (4.2.2) we have

$$\begin{aligned} \dot{F} &= R(\dot{U} + \dot{\theta}AU), \\ \ddot{F} &= R(\ddot{U} + \ddot{\theta}AU + \dot{\theta}^2 A^2 U + 2\dot{\theta}A\dot{U}). \end{aligned} \tag{4.2.5}$$

Moreover, the current internal force-moment Σ has the representation $(2.7.13)_2$, namely,

$$\Sigma = R\mathfrak{S}_u(U)UR^T. \tag{4.2.6}$$

Placing $(4.2.5)_2$ and (4.2.6) into (4.2.3), we obtain the following reduced equation of motion for roto-deformations:

$$(\ddot{U} + \ddot{\theta}AU + \dot{\theta}^2 A^2 U + 2\dot{\theta}A\dot{U})E_R U = R^T MR - \mathfrak{S}_u(U)U. \tag{4.2.7}$$

This equation is general enough to include the treatment of anisotropic elastic bodies with arbitrary principal Euler axes. It does not require that the director basis elements D_i lie along the principal axes of E_R, though specializing assumptions of this kind are often useful in particular cases.

CASE 1 (Steady-state roto-deformations for isotropic materials). We consider first solutions for which

$$\dot{\theta} = \omega, \qquad U = U_0 = \sum_{i=1}^{3} u_i D_{ii}, \tag{4.2.8}$$

ω and u_i being positive constants. The u_i are the principal stretches of the deformation. Note that we are assuming that the axis of rotation D_1 is also a principal axis of stretch. If we further assume that $M = 0$, and that E_R also has D_1 as a principal axis, then (4.2.7) reduces to

$$\omega^2 A^2 U_0 E_R = -\mathfrak{S}_u(U_0), \tag{4.2.9}$$

and E_R must have the form

$$E_R = \sum_{i=1}^{3} E_i D_{ii} + E_{23} D_{23}. \tag{4.2.10}$$

We also note from (4.1.6) and (4.1.7) that

$$A^2 = A_1^2 = -D_{22} - D_{33}. \tag{4.2.11}$$

A final important simplification of (4.2.9) is obtained by focusing on isotropic materials. For these it follows from $(2.7.13)_1$, $(2.7.19)_1$, and (4.1.7) that $\mathfrak{S}_u$ can be expressed in terms of principal force-moments T_i by

$$\mathfrak{S}_u = \sum_{i=1}^{3} T_i D_{ii}. \tag{4.2.12}$$

Moreover, by (2.7.24) the principal force-moments are given as functions of the principal stretches by a single response function τ. Using (4.2.10)–(4.2.12) and (2.7.24) in (4.2.9), we finally arrive at the system of equations

$$\begin{aligned}
\tau(u_1, u_2, u_3) &= 0, \\
\tau(u_2, u_3, u_1) &= \omega^2 u_2 E_2, \\
\tau(u_3, u_1, u_2) &= \omega^2 u_3 E_3, \\
E_{23} &= 0.
\end{aligned} \tag{4.2.13}$$

The last equation shows that U_0 and E_R must have the same principal axes. For ω given, the other three equations present an algebraic problem to be solved for the principal stretches.

As an illustration of the type of information that (4.2.13) can deliver, we consider the special case of a St. Vénant–Kirchhoff material. For it the function τ is given by $(2.7.33)_2$, and so the system (4.2.13) becomes

$$\begin{aligned}
[\lambda(u_1^2 + u_2^2 + u_3^2) + 2\mu u_1^2 - (3\lambda + 2\mu)]u_1 &= 0, \\
[\lambda(u_1^2 + u_2^2 + u_3^2) + 2\mu u_2^2 - (3\lambda + 2\mu)]u_2 &= 2\omega^2 u_2 E_2, \\
[\lambda(u_1^2 + u_2^2 + u_3^2) + 2\mu u_3^2 - (3\lambda + 2\mu)]u_3 &= 2\omega^2 u_3 E_3.
\end{aligned} \tag{4.2.14}$$

This implies that

$$u_1^2 = 1 - \frac{\lambda \omega^2 (E_2 + E_3)}{\mu(3\lambda + 2\mu)},$$

$$u_2^2 = 1 + \frac{E_2 \omega^2}{\mu} - \frac{\lambda \omega^2 (E_2 + E_3)}{\mu(3\lambda + 2\mu)}, \qquad (4.2.15)$$

$$u_3^2 = 1 + \frac{E_3 \omega^2}{\mu} - \frac{\lambda \omega^2 (E_2 + E_3)}{\mu(3\lambda + 2\mu)}.$$

Since μ, $3\lambda + 2\mu$, E_2, E_3, and ω^2 are all positive, all three stretches will exceed 1 if $\lambda < 0$. While we might not expect this physically, there is no restriction in the theory to disallow it for certain materials. When $\lambda > 0$, there are some unusual features of the solution. In this case $u_1^2 < 1$, so a contraction occurs along the axis of spin. However, we must have $u_1^2 > 0$, and this implies that

$$\omega^2 < \frac{\mu(3\lambda + 2\mu)}{\lambda(E_2 + E_3)}. \qquad (4.2.16)$$

We view this as placing a limitation on the theory for the St. Vénant–Kirchhoff material: *the larger the moments of inertia E_2 and E_3, the smaller is the range of allowable spin rate.* That such a limitation should indeed arise is evident from $(2.7.33)_1$. In any reasonable material we may expect the principal force-moments to increase without bound as $\det \mathbf{F} \to 0$, that is, as $u_1 u_2 u_3 \to 0$, but $(2.7.33)_1$ does not exhibit this behavior.

Rearranging the solution slightly, we note that u_2^2 and u_3^2 can be written as

$$u_2^2 = 1 + \frac{E_2 \omega^2 \lambda}{\mu(3\lambda + 2\mu)} \left(\frac{2(\lambda + \mu)}{\lambda} - \frac{E_3}{E_2} \right),$$

$$u_3^2 = 1 + \frac{2E_2 \omega^2 (\lambda + \mu)}{\mu(3\lambda + 2\mu)} \left(\frac{E_3}{E_2} - \frac{\lambda}{2(\lambda + \mu)} \right). \qquad (4.2.17)$$

Since $\lambda > 0$ and $\mu > 0$, we see that $2(\lambda + \mu)/\lambda > 2$ and $\lambda/2(\lambda + \mu) = (\lambda + \mu - \mu)/2(\lambda + \mu) = \frac{1}{2} - \mu/2(\lambda + \mu) < \frac{1}{2}$. Therefore, (4.2.17) shows that

$$u_2^2 > 1 \quad \text{and} \quad u_3^2 < 1 \quad \text{when} \quad \frac{E_3}{E_2} < \frac{\lambda}{2(\lambda + \mu)},$$

$$u_2^2 > 1 \quad \text{and} \quad u_3^2 > 1 \quad \text{when} \quad \frac{\lambda}{2(\lambda + \mu)} < \frac{E_3}{E_2} < \frac{2(\lambda + \mu)}{\lambda}, \qquad (4.2.18)$$

$$u_2^2 < 1 \quad \text{and} \quad u_3^2 > 1 \quad \text{when} \quad \frac{E_3}{E_2} > \frac{2(\lambda + \mu)}{\lambda}.$$

Therefore, only when E_2 and E_3 are comparable in size will both principal stretches in the plane of rotation exceed 1. If the body is quite asymmetric, so E_2 is considerably larger than E_3 or E_3 is considerably larger than E_2, there will be an extension in the direction corresponding to the larger moment of inertia and a contraction in the direction corresponding to the smaller.

CASE 2 (Axially symmetric unsteady roto-deformations for isotropic materials with an axis of symmetry). Consider next solutions of (4.2.7) in which both E_R and U have a common axis of symmetry coincident with the axis of spin D_1:

$$E_R = E_1 D_{11} + E(D_{22} + D_{33}),$$
$$U = u D_{11} + v(D_{22} + D_{33}). \tag{4.2.19}$$

E_1 and E are positive constants while u and v are positive functions of time. If we also restrict the force-moment M to be a compression or tension along the axis of spin, possibly time dependent, then

$$M = M(t) D_{11}. \tag{4.2.20}$$

Since D_1 is the axis of spin, $R^T M R = M$, and therefore

$$R^T M R U^{-1} E_R^{-1} = M D_{11} \left(\frac{1}{E_1 u} D_{11} + \frac{1}{Ev}(D_{22} + D_{33}) \right)$$

$$= \frac{M}{E_1 u} D_{11}. \tag{4.2.21}$$

For an isotropic material $\mathfrak{S}_u(U)$ is given by (4.2.12) and (2.7.24), and then (4.2.19) shows that

$$\mathfrak{S}_u(U) E_R^{-1} = \frac{1}{E_1} \tau(u, v, v) D_{11} + \frac{1}{E} \tau(v, v, u)(D_{22} + D_{33}). \tag{4.2.22}$$

Finally, we note from $(4.2.19)_2$, (4.1.6), (4.1.7), and (4.2.11) that

$$\ddot{U} + \dot{\theta}^2 A^2 U = \ddot{u} D_{11} + (\ddot{v} - v\dot{\theta}^2)(D_{22} + D_{33}),$$
$$\ddot{\theta} A U + 2\dot{\theta} A \dot{U} = (v\ddot{\theta} + 2\dot{v}\dot{\theta}) A_1. \tag{4.2.23}$$

Therefore, (4.2.7) reduces in view of (4.2.21)–(4.2.23) to

$$E_1 u \ddot{u} = M(t) - u\tau(u, v, v),$$
$$E(\ddot{v} - v\dot{\theta}^2) = -\tau(v, v, u), \tag{4.2.24}$$
$$v\ddot{\theta} + 2\dot{v}\dot{\theta} = 0.$$

The last equation here can be integrated explicitly to give

$$v^2 \dot{\theta} = A, \tag{4.2.25}$$

where A is constant. This expresses the conservation of angular momentum for the solutions we are considering. Using this result to eliminate $\dot{\theta}$, we obtain the final system

$$E_1 u\ddot{u} = M(t) - u\tau(u, v, v),$$
$$Ev^3\ddot{v} = EA^2 - v^3\tau(v, v, u).$$

(4.2.26)

First set $M = 0$. We are left with a second-order system for the principal stretches u and v, given the angular momentum A. Steady-state solutions satisfy $\dot{u} = \dot{v} = 0$, and hence

$$\tau(u, v, v) = 0,$$
$$v^3\tau(v, v, u) = EA^2.$$

(4.2.27)

If the unstrained state is natural, that is, $\tau(1, 1, 1) = 0$, then (4.2.27) is satisfied when $u = v = 1$ and $A = 0$. If we assume that

$$\det(\partial T_i/\partial v_j) \neq 0,$$

(4.2.28)

then a detailed calculation and application of the implicit function theorem shows that for A sufficiently small (4.2.27) has a unique solution

$$u = \bar{u}(EA^2), \qquad v = \bar{v}(EA^2)$$

(4.2.29)

satisfying $\bar{u}(0) = \bar{v}(0) = 1$. The restriction (4.2.28) on the function τ is precisely by IFS^+ condition[4] ensuring local invertibility of the principal force-moment–stretch relations (2.7.24).

In order to examine the stability of this steady state, we set

$$u = \bar{u}(EA^2) + r, \qquad v = \bar{v}(EA^2) + s$$

(4.2.30)

and then linearize (4.2.26) in r and s. As a result, we obtain the linear differential system

$$\ddot{r} = a_{11}(EA^2)r + a_{12}(EA^2)s,$$
$$\ddot{s} = a_{21}(EA^2)r + a_{22}(EA^2)s,$$

(4.2.31)

where the functions a_{ij} are defined and smooth for A near zero. When $A = 0$, these equations represent a linearization about a natural state for an isotropic material, and in this case one can show[5] that only oscillations are possible. Moreover, this conclusion is not sensitive to small perturbations of A, and so oscillations also occur for (4.2.31) for all A in a neighborhood of zero. Furthermore, if the governing constitutive relations are semilinear (cf. (2.7.29)), then (4.2.26) gives rise to oscillations for all A.

We turn now to a situation in which the nonlinear problem (4.2.26) can be analyzed directly. Assume $u \equiv 1$ and choose the force-moment component $M(t)$ so that $(4.2.26)_1$ is satisfied. This is a constrained problem in which no stretch along the axis of spin is permitted. We are left then

---

[4] TRUESDELL and NOLL [1965, Eq. (51.14)].

[5] COHEN and MUNCASTER [1984, §4].

with $(4.2.26)_2$, namely,

$$\ddot{v} = \frac{A^2}{v^3} - \frac{1}{E}\tau(v, v, 1).$$ (4.2.32)

This can be integrated once to give

$$\dot{v}^2 + \frac{A^2}{v^2} + \int_1^v \frac{2}{E}\tau(s, s, 1)\, ds = C,$$ (4.2.33)

and now the analysis of solutions can be carried out by phase portrait methods once τ is known. For example, for a St. Vénant–Kirchhoff material we see from $(2.7.33)_2$ that $\tau(s, s, 1) = (\lambda + \mu)(s^2 - 1)s$, and therefore (4.2.33) becomes

$$\dot{v}^2 + \frac{A^2}{v^2} + \frac{\lambda + \mu}{2E}(v^2 - 1)^2 = C.$$ (4.2.34)

For the value $A^2 = (\lambda + \mu)/2E$ and various values of C, the portrait is shown in Fig. 1. Typically, as in this case, the portrait consists of closed curves and so oscillations occur even in the nonlinear problem. The portrait is qualitatively similar to that for $A = 0$ except that it has been moved to the right. There is a single steady state or equilibrium with $v > 1$.

CASE 3 (The bearing problem). Let PP' and QQ' be two rigid, frictionless, parallel plates. Consider a pseudo-rigid ellipsoid spinning about its axis

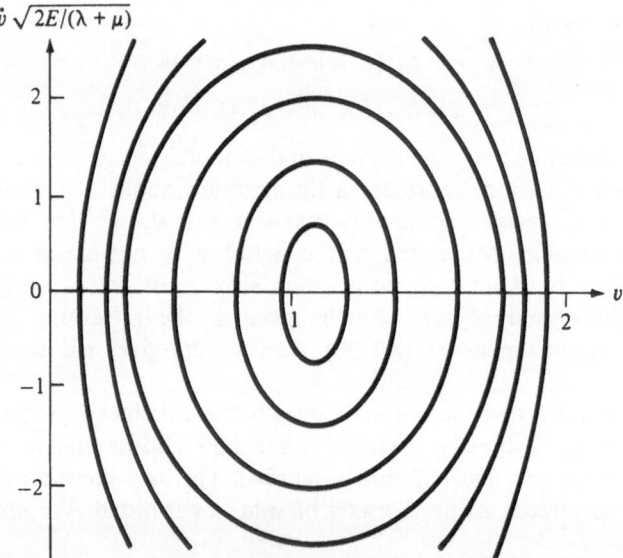

Figure 1

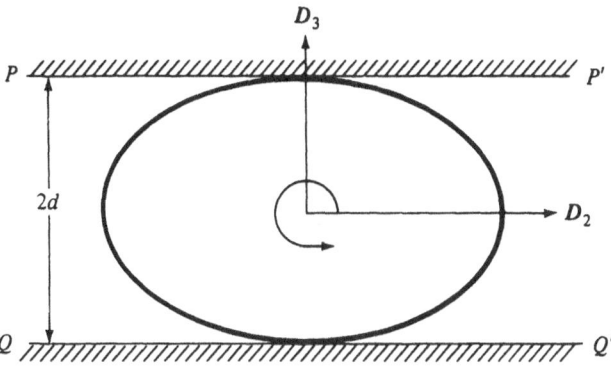

Figure 2

of symmetry D_1, which is located midway between and parallel to the plates (cf. Fig. 2). The circular section of the body normal to the axis of spin has radius a in the reference placement. The dimensions of the body during its motion are such that the plates exert a force-moment

$$M = M(t)D_{33} \qquad (4.2.35)$$

on the body. The objective is to determine the constant value of M for which the body can be in steady motion with constant spin

$$\dot{\theta} = \omega = \text{const.} \qquad (4.2.36)$$

We call this the *bearing problem*.

In this type of deformation the body rotates, but the strains rotate by an equivalent amount so that the final profile appears fixed in space. In terms of the polar decomposition of F we take U to be

$$U = R^{\mathrm{T}} V_0 R, \qquad (4.2.37)$$

where V_0 is a constant positive-definite and symmetric tensor of the form

$$V_0 = \sum_{i=1}^{3} u_i D_{ii} + 2u_{23} D_{23}. \qquad (4.2.38)$$

R is given again by (4.2.2). Equivalently, we consider deformations whose left stretch tensor V has the constant value V_0. Since the body has D_1 as an axis of symmetry, E_R is given by (4.2.19)$_1$, and therefore E_R and R commute. By (4.2.36), (4.2.37), and (4.2.1) we have

$$F = V_0 R, \qquad \dot{F} = \omega V_0 A R, \qquad \ddot{F} = \omega^2 V_0 A^2 R, \qquad (4.2.39)$$

and so

$$\ddot{F} E_R F^{\mathrm{T}} = \omega^2 V_0 A^2 R E_R R^{\mathrm{T}} V_0 = \omega^2 V_0 A^2 E_R V_0. \qquad (4.2.40)$$

For an isotropic material, (4.2.4)$_2$ and (2.7.14)$_2$ show that

$$\Sigma = \mathfrak{S}(F) = \mathfrak{S}_v(V_0)V_0 = \mathfrak{S}(V_0). \qquad (4.2.41)$$

Therefore, by $(2.7.15)_1$, (4.2.40), and (4.2.41), the basic equation (4.2.3) becomes

$$\omega^2 V_0 A^2 E_R V_0 = M - Q_0 I - Q_1 V_0^2 - Q_2 V_0^4. \qquad (4.2.42)$$

We can express this in component form by use of (4.2.35), (4.2.11), $(4.2.19)_1$, and (4.2.38). The calculation is lengthy but straightforward, and it yields the system

$$
\begin{aligned}
0 &= Q_0 + Q_1 u_1^2 + Q_2 u_1^4, \\
E(u_2^2 + u_{23}^2)\omega^2 &= Q_0 + Q_1(u_2^2 + u_{23}^2) \\
&\quad + Q_2((u_2^2 + u_{23}^2)^2 + u_{23}^2(u_2 + u_3)^2), \\
E(u_3^2 + u_{23}^2)\omega^2 &= Q_0 + Q_1(u_3^2 + u_{23}^2) \\
&\quad + Q_2((u_3^2 + u_{23}^2)^2 + u_{23}^2(u_2 + u_3)^2) - M, \\
Eu_{23}(u_2 + u_3)\omega^2 &= Q_1 u_{23}(u_2 + u_3) + Q_2 u_{23}(u_2 + u_3)(u_2^2 + u_3^2 + 2u_{23}^2).
\end{aligned}
\qquad (4.2.43)
$$

Consistent with this system are the special values

$$u_3 = \frac{d}{a}, \qquad u_{23} = 0, \qquad (4.2.44)$$

which would appear appropriate for the bearing problem. Then $(4.2.43)_3$ becomes the formula

$$M = Q_0 + (Q_1 - E\omega^2)\frac{d^2}{a^2} + Q_2\frac{d^4}{a^4}, \qquad (4.2.45)$$

and for the determination of u_1 and u_2 we have

$$
\begin{aligned}
Q_0 + Q_1 u_1^2 + Q_2 u_1^4 &= 0, \\
Q_0 + (Q_1 - E\omega^2)u_2^2 + Q_2 u_2^4 &= 0.
\end{aligned}
\qquad (4.2.46)
$$

If the unstretched state of the material is natural and the invertibility criterion (4.2.28) applies, then the implicit function theorem shows that a unique solution of (4.2.46) exists for all ω^2 near 0 and d/a near 1.

For specific materials more information can be derived from (4.2.46). For example, for the St. Vénant–Kirchhoff material the coefficients Q_1, Q_2, and Q_3 are given by $(2.7.33)_{3-5}$, and then (4.2.46) becomes

$$
\begin{aligned}
\lambda(u_1^2 + u_2^2 + d^2/a^2) - (3\lambda + 2\mu) + 2\mu u_1^2 &= 0, \\
\lambda(u_1^2 + u_2^2 + d^2/a^2) - (3\lambda + 2\mu) + 2\mu u_2^2 - 2E\omega^2 &= 0.
\end{aligned}
\qquad (4.2.47)
$$

These equations are easily solved for u_1 and u_2, and then the formula for M can be written explicitly. The results are

$$u_1^2 = 1 + \frac{\lambda}{2(\lambda + \mu)} \left(1 - \frac{d^2}{a^2} - \frac{E\omega^2}{\mu} \right),$$

$$u_2^2 = 1 + \frac{\lambda}{2(\lambda + \mu)} \left(1 - \frac{d^2}{a^2} + \frac{(\lambda + 2\mu)E\omega^2}{\mu\lambda} \right), \qquad (4.2.48)$$

$$M = -\frac{\mu(3\lambda + 2\mu)}{2(\lambda + \mu)} \frac{d^2}{a^2} \left(1 - \frac{d^2}{a^2} + \frac{(\lambda + 2\mu)E\omega^2}{\mu(3\lambda + 2\mu)} \right).$$

Since $d/a < 1$ and $E > 0$, we see that $M < 0$. Thus, the external force-moment required is always a compression, increasing in magnitude as the spin rate ω increases. For small spins ($\omega \ll 1$) or small moments of inertia ($E \ll 1$), we note that u_1 and u_2 are nearly equal and greater than 1. This is somewhat unexpected because one is a stretch along the axis of rotation, while the other is a stretch in the plane of rotation. The reason for this behavior can be seen most easily in terms of an intermediate deformation of the body. If $\omega = 0$, the above solution represents the final static placement of a circular body compressed between two plates. For this configuration we may certainly expect both u_1 and u_2 to exceed 1. *From this intermediate placement*, (4.2.48) shows that the effect of rotation is then to decrease u_1 and to increase u_2, a behavior that is also not unexpected. The solution (4.2.48) also points out in various ways the limitations of the St. Vénant–Kirchhoff material. For large spin rates we see that u_1^2 becomes negative, and therefore a restriction similar to (4.2.16) is necessary. Also, as the gap separation decreases from $d = a$ to $d = 0$, the compressive force-moment M initially increases as expected, but eventually it reaches a maximum and thereafter decreases to 0.

(iii) Pure Stretch Motions

Within the spectrum of motions of a pseudo-rigid body, rigid motions lie at one extreme; we turn now to solutions that lie at the other. Here we assume that there is no rigid motion; that is, $R \equiv I$ and the center of mass is fixed. Then the polar decomposition of F becomes

$$F = U, \qquad (4.3.1)$$

and we seek solutions of (4.2.3) in this form. These are called *pure stretch motions* of a pseudo-rigid body. Substituting (4.3.1) into (4.2.3) and using $(2.7.13)_2$, we obtain the basic differential equation governing such motions in the form

$$\ddot{U} = MU^{-1}E_R^{-1} - \mathfrak{S}_u(U)E_R^{-1}. \qquad (4.3.2)$$

As in the previous sections, we confine attention to a number of special cases.

CASE 1 (Oscillations of a symmetric and semilinear elastic body). We begin with a completely symmetric body,

$$E_R = EI, \tag{4.3.3}$$

and assume that this body is free of external loads:

$$M = 0. \tag{4.3.4}$$

Let the undeformed state of the body be natural, and assume that the material comprising the body is governed by a semilinear constitutive law, namely $(2.7.29)_3$. With these specializations, the basic equation of motion (4.3.2) reduces to

$$\ddot{S} + \bar{\lambda}(\operatorname{tr} S)I + 2\bar{\mu}S = 0, \tag{4.3.5}$$

where

$$S = U - I, \qquad \bar{\lambda} = \lambda/E, \qquad \bar{\mu} = \mu/E. \tag{4.3.6}$$

S is called the *strain tensor* of the body.

In order to solve (4.3.5), we first take the trace:

$$(\operatorname{tr} S)\ddot{} + (3\bar{\lambda} + 2\bar{\mu}) \operatorname{tr} S = 0. \tag{4.3.7}$$

The general solution of this equation is

$$\operatorname{tr} S = A \cos \omega t + B \sin \omega t, \qquad \omega = (3\bar{\lambda} + 2\bar{\mu})^{1/2}, \tag{4.3.8}$$

where A and B are constants determined by the initial conditions. Substituting this into (4.3.5), we obtain the linear nonhomogeneous differential equation

$$\ddot{S} + 2\bar{\mu}S = -\bar{\lambda}(A \cos \omega t + B \sin \omega t)I. \tag{4.3.9}$$

The general solution has the form

$$S = S_H + S_P, \tag{4.3.10}$$

where S_H is the general solution of the homogeneous equation and S_P is any particular solution of (4.3.9). Elementary analysis shows that S_H and S_P are given, respectively, by

$$S_H = \cos \Omega t\, A + \sin \Omega t\, B, \qquad \Omega = (2\bar{\mu})^{1/2},$$
$$S_P = \tfrac{1}{3}(A \cos \omega t + B \sin \omega t)I. \tag{4.3.11}$$

The arbitrary constant tensors A and B are determined from the initial conditions of the problem, namely, $U(0)$ and $\dot{U}(0)$. We note that the deformation exhibits *two frequencies of oscillation*. Terms with the frequency ω represent an overall dilatation of the body, while terms with the frequency Ω represent variations from the dilatation produced by unequal initial principal stretches or rates of stretch.

CASE 2 (Dilatation of symmetric elastic bodies). As in Case 1, we assume the body is completely symmetric and the external loads vanish. Then (4.3.2) can be written as

$$\ddot{U} + E^{-1}\mathfrak{S}_u(U) = 0. \tag{4.3.12}$$

We seek solutions of (4.3.12) that are *finite dilatations*; that is,

$$U = uI. \tag{4.3.13}$$

In this case, the tensor-valued differential equation (4.3.12) reduces by (4.2.12) and (2.7.24) to the single scalar equation

$$\ddot{u} = -E^{-1}\tau(u, u, u) = -E^{-1}T(u), \tag{4.3.14}$$

where $T(u)$ is the common principal force-moment. A first integral of this equation is given by

$$\dot{u}^2 + 2E^{-1}\int_1^u T(s)\, ds = C, \tag{4.3.15}$$

where C is a constant of integration. Utilizing (4.3.15), we can now analyze solutions with phase portrait methods for chosen forms of the response function T.

For the special case of a St. Vénant–Kirchhoff material, $(2.7.33)_2$ shows that

$$2T(s) = (3\lambda + 2\mu)(s^2 - 1)s, \tag{4.3.16}$$

and therefore (4.3.15) becomes

$$\dot{u}^2 + \tfrac{1}{4}(3\lambda + 2\mu)(u^2 - 1)^2 E^{-1} = C. \tag{4.3.17}$$

For u near 1, the portrait is very similar to that shown in Fig. 1, and this indicates radial oscillations. However, for u near 0, closed curves can intersect the axis $u = 0$ and so violate the restriction $u > 0$. This behavior is simply another reflection of the inappropriateness of the St. Vénant–Kirchhoff material for phenomena involving large compressions.

We can turn the analysis of (4.3.15) around and ask for reasonable restrictions on the response function τ ensuring that u remains positive. One common restriction, based on physical arguments, is that the common principal force-moment $T(u) = \tau(u, u, u)$ should satisfy

$$T(1) = 0,$$

$$T(u) < 0 \text{ for } u < 1, \qquad T(u) \to -\infty \text{ as } u \to 0, \tag{4.3.18}$$

$$T(u) > 0 \text{ for } u > 1, \qquad T(u) \to +\infty \text{ as } u \to +\infty.$$

If we write (4.3.15) in the form

$$\dot{u}^2 = C + 2E^{-1}\int_u^1 T(s)\, ds \tag{4.3.19}$$

and appeal to $(4.3.18)_2$, we see that the integral appearing here will be negative for all $u < 1$. No curves in the phase plane will cross the $\dot{u}$ axis if the right-hand side of (4.3.19) is negative for all C and for all u sufficiently near 0. This means we must require that

$$\lim_{u \to 0} \int_u^1 T(s)\, ds = -\infty. \tag{4.3.20}$$

This behavior is not implied by $(4.3.18)_2$ as the simple choice $T(u) = u - u^{-1/2}$ shows. The condition (4.3.20) is perhaps more transparent when expressed in terms of the stored-energy function

$$\sigma(u) = \int_1^u T(s)\, ds; \tag{4.3.21}$$

for it (4.3.18) and (4.3.20) imply that

$$\sigma(1) = 0, \qquad \sigma(u) > 0 \quad \text{for } u \neq 1,$$
$$\sigma(u) \to \infty \text{ as } u \to 0 \text{ or } +\infty. \tag{4.3.22}$$

If in addition σ is a convex function of u, as is often assumed, the phase portrait for (4.3.15) is qualitatively similar to Fig. 1, consisting entirely of closed curves encircling a single equilibrium point at $u = 1$.

CASE 3 (Rotating pure stretch of an axially symmetric body). This problem is an interesting reinterpretation of the bearing problem discussed in Section 4(ii). Recall that in the bearing problem we considered a body rotating between the plates PP' and QQ' in Fig. 2, which were fixed in space. Here we consider exactly the opposite; namely, the body is fixed in space and the plates rotate relative to it while maintaining a fixed separation. This causes the deformation to float around the axis of rotation, and the body appears to be rotating with the plates while in fact it is undergoing only a pure stretch that is time dependent.

As in the bearing problem, E_R is given by $(4.2.19)_1$, R by (4.2.2) with $\dot{\theta} = \omega = \text{const.}$, and we let V_0 be a constant positive-definite and symmetric tensor of the form (4.2.38). In place of $(4.2.39)_1$, we now consider motions of the type

$$F = R V_0 R^\mathsf{T}. \tag{4.3.23}$$

Note that F is indeed a pure stretch; the orthogonal transformation in its polar decomposition is now I rather than R as in the bearing problem. In this case, neither the left nor the right stretch tensor is constant. Both are equal to the right-hand side of (4.3.23). Writing $F = U$ we see that

$$\dot{U} = \omega R (A V_0 - V_0 A) R^\mathsf{T},$$
$$\ddot{U} = \omega^2 R (A^2 V_0 + V_0 A^2 - 2 A V_0 A) R^\mathsf{T}, \tag{4.3.24}$$

and therefore, since E_R and R commute,

$$\ddot{U}E_R U^T = \omega^2 R(A^2 V_0 + V_0 A^2 - 2A V_0 A)E_R V_0 R^T. \qquad (4.3.25)$$

Note that only the second of the three terms appearing here played a role in the bearing problem (cf. Eq. $(4.2.40)_2$).

Considering once again an isotropic material, we see from $(2.7.13)_2$ (with $R = I$) and $(2.7.15)_1$ that the basic equation (4.3.2) takes the form

$$\omega^2(A^2 V_0 + V_0 A^2 - 2A V_0 A)E_R V_0 = R^T MR - Q_0 I - Q_1 V_0^2 - Q_2 V_0^4. \qquad (4.3.26)$$

Necessarily, $R^T MR$ is independent of time. In this respect it is important to note that M is the force-moment in the present placement. Since the plates themselves are rotating with constant angular velocity ω, we expect M to correspond to a compressive force of constant strength in the direction $d_3 = RD_3$; that is,

$$M = M d_3 \otimes d_3. \qquad (4.3.27)$$

The preceding equations of motion are equivalent to four independent algebraic equations in the unknowns u_1, u_2, u_{23}, and M. The third stretch u_3 is determined by the geometric constraint $(4.2.44)_1$. To obtain these component equations, we substitute (4.2.2), (4.2.11), $(4.2.19)_1$, (4.2.38), and (4.3.27) into (4.3.26) and resolve into components with respect to the tensor basis induced by D_i. As in the bearing problem, we take $u_{23} = 0$. We find that M has the constant value given by

$$M = Q_0 + 2E\omega^2 u_2(d/a) + (Q_1 - 2E\omega^2)(d/a)^2 + Q_2(d/a)^4, \qquad (4.3.28)$$

while the principal stretches u_1 and u_2 are determined by the algebraic system

$$Q_0 + Q_1 u_1^2 + Q_2 u_1^4 = 0,$$
$$Q_0 + (Q_1 - 2E\omega^2)u_2^2 + Q_2 u_2^4 + 2E\omega^2 u_2(d/a) = 0. \qquad (4.3.29)$$

Equations (4.3.28) and (4.3.29) could certainly be specialized to a St. Vénant–Kirchhoff material, but the algebraic system that results does not admit a simple exact solution and so we do not pursue the analysis further.

(iv) Shearing Motions

One of the more important homogeneous deformations treated in mechanics is that referred to as *simple shear*. Because it is theoretically simple and experimentally producible, it has formed the subject of a wealth of studies. A *simple shearing motion* of a pseudo-rigid body is defined by the equation

$$F = I + k(t)D_1 \otimes D_2, \qquad (4.4.1)$$

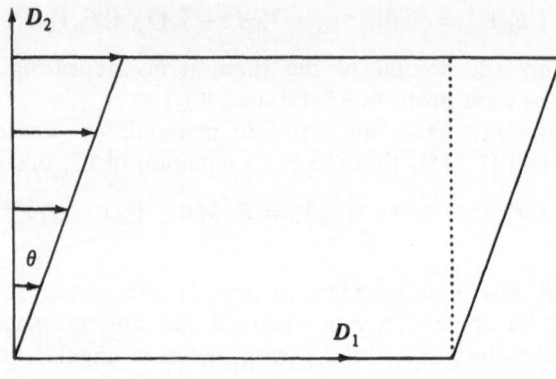

Figure 3

where the D_i, $i = 1$, 2, 3, form a right-handed orthonormal system. The shear is said to be in the direction D_1, of amount $k(t)$, where $k = \tan \theta$ and θ is the angle of shear illustrated in Fig. 3. Simple shear differs from the solutions we have considered so far in two respects. First, there is no particular advantage gained in splitting F into orientation and pure stretch parts by the polar decomposition (4.2.1). A direct analysis of F is more appropriate. Second, a nontrivial force-moment is required to produce simple shear, and some care is necessary in the interpretation of the components of M in order to achieve *free shear oscillations*.

From (4.4.1) and (2.7.16) we find that

$$\ddot{F} = \ddot{k} D_1 \otimes D_2, \qquad F^T = I + k D_2 \otimes D_1,$$

$$FF^T = I + k^2 D_{11} + 2k D_{12},$$

$$(FF^T)^2 = I + k^2(3 + k^2) D_{11} + k^2 D_{22} + 2k(2 + k^2) D_{12},$$

$$I = II = 3 + k^2, \qquad III = 1.$$

$$(4.4.2)$$

If initially we allow a general Euler tensor

$$E_R = E_{ij} D_{ij}, \qquad E_{ij} = E_{ji}, \tag{4.4.3}$$

then

$$\ddot{F} E_R F^T = \ddot{k}[(k E_{22} + E_{21}) D_{11} + E_{22} D_1 \otimes D_2 + E_{23} D_1 \otimes D_3]. \tag{4.4.4}$$

Next we note that for an isotropic material $(2.7.15)_1$ and $(4.4.2)_{3,4}$ give the constitutive equation

$$\mathfrak{S}(F) = [n_0 + n_1(1 + k^2) + n_2(1 + k^2(3 + k^2))] D_{11}$$

$$+ [n_0 + n_1 + n_2(1 + k^2)] D_{22} + [n_0 + n_1 + n_2] D_{33}$$

$$+ 2[n_1 + n_2(2 + k^2)] k D_{12}, \tag{4.4.5}$$

where the response functions n_0, n_1, and n_2 are functions only of the shear parameter k; namely,

$$n_i(k) = Q_i(3 + k^2, 3 + k^2, 1), \qquad i = 0, 1, 2. \tag{4.4.6}$$

The motion is entirely prescribed by (4.4.1) if $k(t)$ is known. From this perspective, the basic equation (4.2.3) may be viewed as giving a formula for the force-moment M required to sustain such motion. If we write

$$M = M_{ij} D_i \otimes D_j \tag{4.4.7}$$

and insert (4.4.4) and (4.4.5) into (4.2.3), we obtain the following components for M:

$$M_{11} = \ddot{k}(kE_{22} + E_{21}) + n_0 + n_1(1 + k^2) + n_2[1 + k^2(3 + k^2)],$$

$$M_{22} = n_0 + n_1 + n_2(1 + k^2), \qquad M_{33} = n_0 + n_1 + n_2,$$

$$M_{12} = E_{22}\ddot{k} + k[n_1 + n_2(2 + k^2)], \qquad M_{21} = k[n_1 + n_2(2 + k^2)], \tag{4.4.8}$$

$$M_{13} = E_{23}\ddot{k}, \qquad M_{31} = M_{23} = M_{32} = 0.$$

These prescribe the required components of force-moment, attained perhaps through suitably constructed loading devices.

In Section 2(iv) we examined in some detail physical interpretations of the components M_{ij} of the current force-moment M. In brief, the components $M_{(ij)}$ of sym M represent the action of a double force without torque; each diagonal component $M_{(ii)}$ (no sum) is a tensile or compressive force along the coordinate axis D_i, while each off-diagonal component $M_{(ij)}$ ($i \neq j$) is a shearing force in the plane of D_i and D_j. Moreover, the component $M_{[ij]}$ of sk M represents an applied torque acting along an axis perpendicular to the plane of D_i and D_j. For the shearing problem at hand, it is reasonable to assume that the external loads do not contribute shear forces or torques perpendicular to the plane of shear. Based on the above interpretations, this means that

$$M_{(13)} = M_{(23)} = 0,$$
$$M_{[13]} = M_{[23]} = 0, \tag{4.4.9}$$

or equivalently,

$$M_{13} = M_{31} = M_{23} = M_{32} = 0. \tag{4.4.10}$$

These are compatible with $(4.4.8)_{6-9}$ provided that

$$E_{23} = 0, \tag{4.4.11}$$

a reasonable restriction on the symmetry of the body for (4.4.10) to apply.

Using $(4.4.8)_{1-5}$, we can examine various combinations of the components of M that have special significance. For example, the reaction torque in the plane of D_1 and D_2 necessary to sustain simple shearing follows immediately from the difference of $(4.4.8)_4$ and $(4.4.8)_5$:

$$M_{[12]} = \tfrac{1}{2}\ddot{k}E_{22}. \tag{4.4.12}$$

In particular, *the required torque is zero if and only if* $\ddot{k} = 0$.

A related but much more interesting problem arises from the additional assumption

$$M_{(12)} = 0. \tag{4.4.13}$$

Since $M_{(12)}$ contributes a shear force in the plane of shear, we may regard (4.4.13) as the condition that the shearing motion be *free* or unforced. The equation governing force-free motions derives from the sum of $(4.4.8)_4$ and $(4.4.8)_5$, namely,

$$E_{22}\ddot{k} + 2k[n_1 + n_2(2 + k^2)] = 0. \tag{4.4.14}$$

Multiplying each side by $2\dot{k}$ and integrating, we obtain

$$E_{22}\dot{k}^2 + 4\int_0^k [n_1(s) + n_2(s)(2 + s^2)]s\,ds = C, \tag{4.4.15}$$

and now force-free shearing motions can be investigated by means of phase portrait methods.

As an illustration of the analysis of (4.4.15), we consider the special case of a St. Vénant–Kirchhoff material. By (4.4.6) and $(2.7.33)_{3-5}$, the response coefficients n_i are

$$n_0(k) = 0, \qquad n_1(k) = -\mu + \tfrac{1}{2}\lambda k^2, \qquad n_2(k) = \mu, \tag{4.4.16}$$

and so (4.4.15) becomes

$$E_{22}\dot{k}^2 + \frac{\lambda + 2\mu}{2}k^4 + 2\mu k^2 = C. \tag{4.4.17}$$

The classical inequalities (2.7.28) ensure that the coefficients of k^2 and k^4 here are both positive, and so the phase portrait consists only of closed curves, symmetric about the k axis, encircling the single fixed point $k = 0$. Hence, *the solutions are pure oscillations*.

(v) Plane Motions

Two-dimensional problems in rigid-body mechanics constitute an interesting and self-contained subset of that theory. Such problems arise when all particles in the body move parallel to a fixed plane $\mathscr{P}$ called the *plane of motion*. Moreover, such motions persist entirely under the action of resultant forces and torques that are vectors parallel and perpendicular to $\mathscr{P}$, respectively. Often one pictures the motion of the body as that of a plane figure contained in and moving about $\mathscr{P}$. This figure is usually called a *lamina*.

We adopt the foregoing point of view in studying two-dimensional

motions of pseudo-rigid bodies. This means that the body is to be regarded as a deformable lamina moving in a fixed plane. We refer to
pseudo-rigid motions defined in this way as *plane motions*.

From the perspective of material response, constitutive theory appropriate to a pseudo-rigid lamina falls within the framework of *membrane
theory*. The constitutive equations for membranes are simpler than those
that pertain in the three-dimensional case, and this fact gives plane problems the advantage of algebraic simplicity without sacrificing the essential
features of the motion. When a more detailed analysis is required,
we need only treat two-dimensional problems using the appropriate
three-dimensional constitutive relations, as was the case previously in
Section 4(ii).

The basic equations that govern plane motions of a pseudo-rigid body
are

$$m\ddot{\boldsymbol{r}} = \boldsymbol{f}, \qquad \dot{\boldsymbol{F}}\boldsymbol{E}_\mathrm{R}\boldsymbol{F}^\mathrm{T} = \boldsymbol{M} - \boldsymbol{\Sigma}. \qquad (4.5.1)$$

The material comprising the body is called *elastic* if

$$\boldsymbol{\Sigma} = Q_0(I, II)\boldsymbol{I} + Q_1(I, II)\boldsymbol{B}, \qquad (4.5.2)$$

where

$$\boldsymbol{B} = \boldsymbol{F}\boldsymbol{F}^\mathrm{T}, \qquad I = \mathrm{tr}\,\boldsymbol{B}, \qquad II = \det\boldsymbol{B}. \qquad (4.5.3)$$

In (4.5.1)–(4.5.3) and henceforth the sans serif font is used to emphasize
the two-dimensional nature of the theory. The constitutive relation (4.5.2)
derives[6] from that of a homogeneous isotropic elastic membrane point
and is the analogue of $(2.7.15)_1$. Equations (4.5.3) define a *strain tensor*[7]
and its principal invariants for this two-dimensional case.

Consider the following plane problem.[8] An elastic lamina, which in its
reference placement is circular of radius r, rolls without slipping on a
horizontal rough line. The force of gravity acts vertically. We wish to
describe the state of motion and deformation representing steady-state
rolling. Here, for the first time, we are addressing a problem in which the
center of mass is not fixed. As a consequence, the motion involves both
translation and deformation, and generally both equations in (4.5.1) will
play a fundamental role. We focus on a symmetrical type of rolling and
content ourselves with examining its steady-state behavior. The study of
accelerated general rolling motions is left as an open problem for the
reader.

Let $\boldsymbol{D}_\alpha$, $\alpha = 1, 2$, be an orthonormal director basis that spans the
translation space of $\mathscr{P}$. Moreover, choose $\boldsymbol{D}_1$ along the line of rolling and
$\boldsymbol{D}_2$ along the vertical (cf. Fig. 4). To a large extent the three-dimensional

---

[6] COHEN and WANG [1984].

[7] Usually $\boldsymbol{B}$ would be called the *left Cauchy–Green strain tensor*.

[8] The following analysis is a slightly expanded version of the discussion in COHEN and
MUNCASTER [1986]. A variety of other plane problems for pseudo-rigid bodies have been
considered by COHEN and MACSITHIGH [1988].

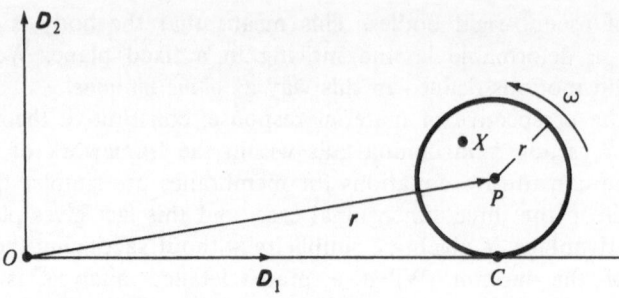

Figure 4

tensor bases given by (4.1.6) and (4.1.7) can easily be adapted to this two-dimensional setting, naturally with certain simplifications and differences. If we set

$$D_{11} = D_1 \otimes D_1, \qquad D_{22} = D_2 \otimes D_2,$$
$$D_{12} = \tfrac{1}{2}(D_2 \otimes D_1 + D_1 \otimes D_2), \qquad (4.5.4)$$
$$A = D_2 \otimes D_1 - D_1 \otimes D_2,$$

then D_{11}, D_{22}, and D_{12} span the space of symmetric tensors while all skew tensors are proportional to A. In particular, we note that

$$A^2 = -I \qquad (4.5.5)$$

in contrast to results such as (4.2.11) for the three-dimensional case.

A general rotation R in the plane $\mathscr{P}$ can be represented in terms of the exponential function following $(4.1.14)_1$, namely,

$$R(t) = \exp \theta(t)A, \qquad \theta(0) = 0. \qquad (4.5.6)$$

The angular velocity tensor is given by

$$W = \dot{R}R^{\mathsf{T}} = \omega A, \qquad (4.5.7)$$

where the *scalar angular velocity* is

$$\omega = \dot{\theta}. \qquad (4.5.8)$$

The effect of this rotation can be viewed more easily in terms of a co-rotating director basis d_α, $\alpha = 1, 2$, given by

$$d_\alpha = RD_\alpha. \qquad (4.5.9)$$

By (4.5.5) and the definition (4.1.8) of the exponential function, the following explicit representation paralleling (4.1.12) can be obtained:

$$R = \cos \theta \, I + \sin \theta \, A. \qquad (4.5.10)$$

Using this and $(4.5.4)_4$ in (4.5.9), we find that

$$d_1 = \cos\theta\,D_1 + \sin\theta\,D_2, \qquad d_2 = -\sin\theta\,D_1 + \cos\theta\,D_2, \quad (4.5.11)$$

which show that θ is indeed the angle of rotation.

CASE 1 (Steady rigid rolling). In order to establish a framework from which to analyze the rolling motion of a pseudo-rigid body, we review first the familiar problem of rigid rolling. Figure 4 illustrates this situation, showing the center of mass P and the point of contact C between the lamina and the line of rolling. X indicates an arbitrary point in the lamina and O is an origin chosen arbitrarily on the plane of rolling. The lamina is assumed to be centered at O in its reference placement κ. Clearly, this leads to a situation in which κ is a placement that the body can never occupy during its motion. If X occupies the place X in κ and the place x at time t in the current placement, then according to (3.3.4) this motion is the affine transformation

$$x = \chi(t, X) = x_0 + r(t) + R(t)(X - x_0), \qquad (4.5.12)$$

where x_0 is the place occupied by the origin O.

For a steady-state condition, we assume that R is given by (4.5.6) in which the scalar angular velocity ω is constant. Then either (4.5.5) or (4.5.10) shows that

$$\ddot{R} = -\omega^2 R. \qquad (4.5.13)$$

The kinematical description is completed by giving formulas for the velocity and acceleration associated with (4.5.12). These are

$$\dot{x} = \dot{r} + \omega A(x - x_0 - r), \qquad \ddot{x} = \ddot{r} - \omega^2(x - x_0 - r). \quad (4.5.14)$$

Into these we must incorporate the requirement that the body roll without slipping. This is embodied in the condition

$$\dot{x}_C = 0, \qquad x_C \equiv x_0 + r - r D_2, \qquad (4.5.15)$$

where x_C is the place occupied by the contact point C at time t. Replacing x by x_C in (4.5.14)$_1$ and using (4.5.15) and (4.5.4)$_4$, we see that

$$\dot{r} = r\omega A D_2 = -r\omega D_1, \qquad \ddot{r} = 0. \qquad (4.5.16)$$

It follows immediately that

$$r = r_0 - r\omega t\,D_1, \qquad r_0 = r(0). \qquad (4.5.17)$$

In this type of problem the equations of motion serve mainly to define the external force and force-moment that are required to produce the rolling motion. The force system acting on the lamina consists of the uniform gravitational body force $b = -\sigma g D_2$ and a reaction N at the contact point C. Here σ and g denote the mass per unit area and acceleration due to gravity, respectively. In rigid-body mechanics, it is customary to write

$$N = N\boldsymbol{D}_1 + F\boldsymbol{D}_2, \tag{4.5.18}$$

where N and F are called the normal and friction reactions, respectively. In addition, an unknown constraint force-moment $\boldsymbol{C}$ must be supplied in order to guarantee the rigidity of the motion. Thus, by calculations similar to those presented in Section 2(iv), we are led to the following expressions for the resultant applied force and force-moment, respectively:

$$\boldsymbol{f} = (N - mg)\boldsymbol{D}_2 + F\boldsymbol{D}_1,$$
$$\boldsymbol{M} = -Nr\boldsymbol{D}_{22} - Fr\boldsymbol{D}_1 \otimes \boldsymbol{D}_2 + \boldsymbol{C}. \tag{4.5.19}$$

Consider finally the equation of motion $(4.5.1)_2$. The Euler tensor for a uniform circular lamina of mass m and radius r is well known. It is given by

$$\boldsymbol{E_R} = \boldsymbol{E} = \tfrac{1}{4}mr^2\boldsymbol{I}. \tag{4.5.20}$$

If we require that the reference placement κ be a natural state, then $\Sigma = \boldsymbol{0}$ when $\boldsymbol{F} = \boldsymbol{I}$. Since $\boldsymbol{B} = \boldsymbol{I}$ both for this case and the case $\boldsymbol{F} = \boldsymbol{R}$, it follows from (4.5.2) that

$$\Sigma = \boldsymbol{0} \tag{4.5.21}$$

in a rigid motion. By (4.5.13), $(4.5.16)_2$, and (4.5.19)–(4.5.21), we conclude from (4.5.1) that the transplacement $(\boldsymbol{r}, \boldsymbol{R})$ for rigid rolling is given by (4.5.10) and (4.5.17) provided that

$$N = mg, \qquad F = 0, \qquad \boldsymbol{C} = mgr\boldsymbol{D}_{22} - \tfrac{1}{4}m\omega^2 r^2 \boldsymbol{I}. \tag{4.5.22}$$

All the foregoing results are standard with the exception of $(4.5.22)_3$. This latter result gives, within the framework of the theory of pseudo-rigid bodies, the constraining force-moment that must be applied to prevent deformation and thereby sustain rigid motion.

Case 2 (Steady symmetrical rolling). Assume now that the lamina is free of the constraining force-moment $\boldsymbol{C}$. We confine attention to motions we call *symmetrical rolling* as illustrated in Fig. 5. This means that the lamina, originally circular, deforms into an ellipse that rolls such that the end of its semi-minor axis is always at the contact point C. The reference placement κ is the same as that in the previous case of rigid rolling. According to (3.3.6), the affine motion $\boldsymbol{x} = \chi(t, \boldsymbol{X})$ induced in this case by the pseudo-rigid motion $(\boldsymbol{r}, \boldsymbol{F})$ is given by

$$\boldsymbol{x} = \chi(t, \boldsymbol{X}) = \boldsymbol{x}_0 + \boldsymbol{r}(t) + \boldsymbol{F}(t)(\boldsymbol{X} - \boldsymbol{x}_0). \tag{4.5.23}$$

The deformation $\boldsymbol{F}$ can be written in terms of its left polar decomposition as

$$\boldsymbol{F} = \boldsymbol{VR}, \tag{4.5.24}$$

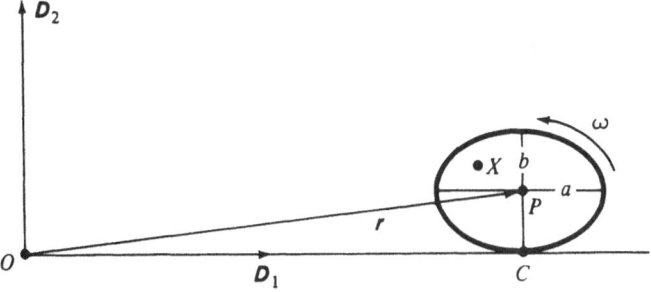

Figure 5

and we assume now that the rotation R is given by (4.5.6) and that the stretch V has the form

$$V = \sum_{\alpha=1}^{2} v_\alpha D_{\alpha\alpha}. \tag{4.5.25}$$

It is useful at this point to include the possibility of an unsteady or oscillatory symmetric rolling by allowing $v_\alpha(t)$ and $\theta(t)$ to be arbitrary functions of t. Later, we shall restriction our attention to *steady-state motion*, which we take to be defined by the condition

$$\ddot{r} \cdot D_1 = 0; \tag{4.5.26}$$

namely, *the mass center is nonaccelerating.*

With (4.5.6), (4.5.24), and (4.5.25), it is straightforward to compute the deforming tensor L defined by (2.2.6). We obtain

$$L = \sum_{\alpha=1}^{2} \left(\frac{\dot{v}_\alpha}{v_\alpha}\right) D_{\alpha\alpha} + \left(\frac{\dot{\theta}}{v_1 v_2}\right)(v_2^2 D_2 \otimes D_1 - v_1^2 D_1 \otimes D_2). \tag{4.5.27}$$

As already mentioned in Section 2(ii), it is common practice in continuum mechanics to decompose L into the sum

$$L = D + O, \tag{4.5.28}$$

where

$$D = \text{sym } L, \qquad O = \text{sk } L. \tag{4.5.29}$$

D and O are the *stretching* and *spin* tensors, respectively, of the motion. From (4.5.27) and (4.5.29), we find that

$$D = \sum_{\alpha=1}^{2} \left(\frac{\dot{v}_\alpha}{v_\alpha}\right) D_{\alpha\alpha} + \dot{\theta}\left[\frac{(v_2^2 - v_1^2)}{v_1 v_2}\right] D_{12},$$

$$O = \dot{\theta}\left[\frac{(v_1^2 + v_2^2)}{2v_1 v_2}\right] A. \tag{4.5.30}$$

More generally, by differentiating (4.5.24) and using (4.5.7)$_1$ and (4.5.25), we can show that

$$D = \dot{V}V^{-1} + \text{sym } VWV^{-1},$$
$$O = \text{sk } VWV^{-1}. \qquad (4.5.31)$$

These equations exhibit the relation between the four tensors D, O, V, and W. Most significant is the fact, well known in continuum mechanics, that the spin O and the angular velocity W are not the same. Of course, for rigid motions the two indeed agree. A second differentiation of (4.5.24) leads to

$$\ddot{F}F^{\mathsf{T}} = \sum_{\alpha=1}^{2} v_\alpha(\ddot{v}_\alpha - v_\alpha\dot{\theta}^2)D_{\alpha\alpha} + \sum_{\alpha,\beta=1}^{2} \varepsilon_{\alpha\beta}v_\alpha(2\dot{v}_\beta\dot{\theta} + v_\beta\ddot{\theta})D_\beta \otimes D_\alpha, \quad (4.5.32)$$

where $\varepsilon_{\alpha\beta}$ is the two-dimensional permutation symbol.

In analogy with (4.5.14) for the case of rigid motion, the velocity and acceleration associated with (4.5.23) are

$$\dot{x} = \dot{r} + L(x - x_0 - r), \qquad \ddot{x} = \ddot{r} + (\dot{L} + L^2)(x - x_0 - r). \quad (4.5.33)$$

If we denote the lengths of the semi-major and semi-minor axes of the ellipse in the current placement by a and b, respectively, then

$$a = rv_1, \qquad b = rv_2. \qquad (4.5.34)$$

The condition for rolling without slipping is still given by (4.5.15)$_1$, but now (4.5.15)$_2$ is replaced by

$$x_C = x_0 + r - bD_2. \qquad (4.5.35)$$

Replacing x by x_C in (4.5.33)$_1$ and using (4.5.15)$_1$, (4.5.27), (4.5.34) and (4.5.35), we find that

$$\dot{r} = -a\dot{\theta}D_1 + r\dot{v}_2 D_2, \qquad \ddot{r} = -(\dot{a}\dot{\theta} + a\ddot{\theta})D_1 + r\ddot{v}_2 D_2. \quad (4.5.36)$$

Note the interesting fact that $\dot{r} \cdot D_1 = -a\dot{\theta}$, not $-b\dot{\theta}$ as might have been guessed intuitively.

The system of applied forces on the lamina arises from gravity and the reaction at C. For completeness, we include also a driving force $D = -DD_1$ acting at P. Thus, the applied resultant force and force-moment are

$$f = (N - mg)D_2 + (F - D)D_1,$$
$$M = -NbD_{22} - FbD_1 \otimes D_2. \qquad (4.5.37)$$

To complete the analysis we require the constitutive law (4.5.2) in component form. To this end we write

$$\Sigma = \sum_{\alpha=1}^{2} T_\alpha(v_1, v_2)D_{\alpha\alpha}, \qquad (4.5.38)$$

where the T_α are the principal force-moments. This is an analogue of (4.2.12) and is a consequence of isotropy and the fact that the v_α are principal stretches. We now substitute $(4.5.36)_2$, the formula (4.5.20) for $\boldsymbol{E}_\mathrm{R}$, (4.5.32), (4.5.34), (4.5.37), and (4.5.38) into (4.5.1) to obtain the differential system

$$mr(\dot{v}_1\dot{\theta} + v_1\ddot{\theta}) = D - F, \qquad mr\ddot{v}_2 = N - mg,$$

$$mr^2 v_1(\ddot{v}_1 - v_1\dot{\theta}^2) = -4T_1(v_1, v_2),$$

$$mr^2 v_2(\ddot{v}_2 - v_2\dot{\theta}^2) = -4T_2(v_1, v_2) - 4Nrv_2, \tag{4.5.39}$$

$$mr^2(2\dot{v}_1\dot{\theta} + v_1\ddot{\theta}) = 4Fr, \qquad mr^2(2\dot{v}_2\dot{\theta} + v_2\ddot{\theta}) = 0.$$

For steady motions we append to this system the requirement (4.5.26), expressed in view of $(4.5.36)_2$ in the form

$$\dot{v}_1\dot{\theta} + v_1\ddot{\theta} = 0. \tag{4.5.40}$$

If in addition we assume that

$$\ddot{\theta} = 0, \qquad \dot{v}_1 = 0, \qquad \dot{v}_2 = 0, \tag{4.5.41}$$

then (4.5.39) and (4.5.40) are satisfied provided that

$$D = F = 0, \qquad N = mg, \tag{4.5.42}$$

and

$$mr^2 v_1^2 \omega^2 = 4T_1(v_1, v_2),$$

$$mr^2 v_2^2 \omega^2 = 4T_2(v_1, v_2) + 4mgrv_2, \tag{4.5.43}$$

where $\omega = \dot{\theta}$ is a constant. Equations (4.5.41)–(4.5.43) comprise sufficient conditions for steady symmetric rolling; (4.5.43) can be solved for the resulting constant stretches. We note that F in this problem is just the projection onto $\mathscr{P}$ of the deformation F occurring in the bearing problem of Section 4(ii). If we now abandon assumption (4.5.41) but suppose that $D = 0$, then from (4.5.39) and (4.5.40) we again conclude that (4.5.41)–(4.5.43) must hold. We summarize the above results as follows.

Theorem. *Let the applied force D parallel to the line of rolling vanish. Then for steady symmetric rolling it is necessary and sufficient that $N = mg$, $F = 0$, $\dot{\theta} = \omega = $ const., and v_1 and v_2 be constants satisfying (4.5.43).*

To illustrate the calculation of v_1 and v_2, consider the linear constitutive relation

$$\Sigma = \varphi_0 I + \varphi_1 \boldsymbol{B}, \tag{4.5.44}$$

where φ_0 and φ_1 are material constants. We have already assumed that the reference placement κ is a natural state, and this implies that (4.5.44) can be rewritten in the form

$$\Sigma = \tfrac{1}{4}\varphi^2 \sum_{\alpha=1}^{2} (v_\alpha^2 - 1) \boldsymbol{D}_{\alpha\alpha}, \qquad (4.5.45)$$

where we have set $\varphi^2 = 4\varphi_1(= -4\varphi_0)$. In this case, (4.5.43) allows solutions of the form

$$v_1 = [1 - mr^2\omega^2/\varphi^2]^{-1/2},$$
$$v_2 = (2mgrv_1/\varphi^2)\{[1 + (\varphi^2/2mgrv_1)^2]^{1/2} - 1\}v_1. \qquad (4.5.46)$$

We note that $v_1 = 1$ when $\omega = 0$, indicating no deformation along the line of rolling, while v_2 in this case reflects the fact that the body has been compressed owing to its own weight. As ω increases, the stretch v_1 increases, and so the body becomes more elongated along the line of rolling.

CHAPTER 5

Variational Formulation of the Theory

Scholars of mechanics generally agree that variational principles provide one of the most elegant, systematic, and satisfying approaches to the study of finite-dimensional dynamical systems. The strength of the variational method lies in its generality. Systematic schemes of approximation, once developed for one problem, are easily adapted to others. Symmetries can be incorporated and exploited in a variety of problems in an essentially uniform and straightforward manner. And deep theorems on the behavior of dynamical systems can readily be applied to a host of otherwise intractable problems.

In order to bring these strengths to bear on the study of pseudo-rigid bodies, we require a variational setting for the theory. Since the equations of motion for a pseudo-rigid body constitute a finite-dimensional dynamical system, it is quite natural to begin with HAMILTON's variational principle and to derive the Lagrangian and Hamiltonian formulations from it. Our objective here is to carry out this development and to use it, together with traditional techniques, to analyze in detail the stability of steady motion and small oscillations for a pseudo-rigid body. A unique feature of the analysis is the fact that the configuration manifold of the variational framework can quite naturally be viewed as a Lie group, and this has a novel impact on the consideration of groups, invariance, symmetry, and conservation laws.

(i) Lagrangian and Hamiltonian Formulations

In the standard variational formulation for a finite dynamical system, the mechanical behavior of the system is assumed to be determined by a real-valued function L called the *Lagrangian* and a real-valued functional A called the *action*. The Lagrangian is a function of the form

$$L = L(q, \dot{q}, t), \tag{5.1.1}$$

where q is a point on the *configuration manifold* $\mathcal{Q}$ of the system, and $\dot{q}$

denotes the velocity vector at q. Hence, the pair $(q, \dot{q})$ is an element in the tangent bundle $T\mathscr{Q}$ based on $\mathscr{Q}$, and so $L: T\mathscr{Q} \times \mathscr{R} \to \mathscr{R}$. The dimension of $\mathscr{Q}$ is assumed to be finite; that is, $\dim \mathscr{Q} = n$. The components of q are called *generalized coordinates* and the system is said to have n *degrees of freedom*. The collection of points $(q, \dot{q})$ is referred to as the *state space* $\mathscr{S}$ for the system. For the theory of pseudo-rigid bodies, it corresponds to the space $\mathscr{S}_p$ introduced in Section 3(i).

Mechanical laws are expressed through a governing variational principle called HAMILTON'S *principle*. If $q = q(t)$ denotes a curve on $\mathscr{Q}$, and if $\mathscr{C}$ denotes the set of such curves having prescribed smoothness, HAMILTON'S principle requires that this curve be a motion of the dynamical system only if it is an extremum of the action. The action functional $A: \mathscr{C} \to \mathscr{R}$ is defined by

$$A[q] = \int_{\mathscr{I}} L(q(t), \dot{q}(t), t), \tag{5.1.2}$$

where the integration is over an interval of time $\mathscr{I} = (t_1, t_2)$ and the measure dt is suppressed. Extremals of A can be determined through standard methods in the calculus of variations. It is well known, for example, that smooth extremals must satisfy the system of Euler–Lagrange equations

$$\frac{d}{dt}\left(\frac{\partial L}{\partial \dot{q}}\right) - \frac{\partial L}{\partial q} = 0. \tag{5.1.3}$$

The Hamiltonian formulation is built on the preceding results and evolves as follows. The *generalized momentum* p conjugate to q is defined by

$$p = \frac{\partial L}{\partial \dot{q}}. \tag{5.1.4}$$

The *canonical variables* q and p are then introduced as defining a point (q, p) on a $2n$-dimensional manifold $\mathscr{P}$ called the *phase space*. Initially, q and p are regarded as independent variables, with $\dot{q}$ being determined implicitly through (5.1.4) as a function of the canonical variables and time: $\dot{q} = Q(q, p, t)$. The *Hamiltonian H* is the real-valued function defined by

$$H(q, p) = p \cdot \dot{q} - L(q, \dot{q}, t)$$
$$= p \cdot Q(q, p, t) - L(q, Q(q, p, t), t). \tag{5.1.5}$$

The generalized momentum p is most naturally viewed as an element of the dual to the tangent space at q, and we write $p(\dot{q}) = p \cdot \dot{q} \in \mathscr{R}$ to denote its value on the vector $\dot{q}$. More precisely, p is an element of the *cotangent space* at q, and $H: T^*\mathscr{Q} \times \mathscr{R} \to \mathscr{R}$, where the cotangent bundle $T^*\mathscr{Q}$ based on $\mathscr{Q}$ serves here as the phase space $\mathscr{P}$. Once again standard variational procedures yield a first-order differential system on $\mathscr{P}$; namely,

$$\dot{q} = \frac{\partial H}{\partial p}, \qquad \dot{p} = -\frac{\partial H}{\partial q}. \tag{5.1.6}$$

These are *HAMILTON's equations of motion*. Their solution curves on the phase space $\mathscr{P}$ are entirely equivalent to curves $q \in \mathscr{C}$ on the configuration space $\mathscr{Q}$ that satisfy the system of Euler–Lagrange equations (5.1.3).

With appropriate choices of the configuration manifold $\mathscr{Q}$ and the Lagrangian L, the preceding structure applies to the theory of pseudo-rigid bodies. We take $\mathscr{Q}$ to be the vector space

$$\mathscr{Q} = \mathscr{V} \oplus \mathscr{K}, \tag{5.1.7}$$

where

$$\mathscr{K} = \bigoplus_{i=1}^{3} \mathscr{V}_i, \tag{5.1.8}$$

and $\mathscr{V}_i = \mathscr{V}$ for $i = 1, 2, 3$. We call the first and second factors of $\mathscr{Q}$ the physical and director components, respectively. Note that dim $\mathscr{Q} = 12$. The generalized coordinates q are defined as ordered quartets of vectors

$$q = (r, d), \qquad d = (d_1, d_2, d_3), \tag{5.1.9}$$

reflecting the decompositions (5.1.7) and (5.1.8). In addition, we assume that the vectors d_i, $i = 1, 2, 3$, form a director basis on each $\mathscr{V}_j$, $j = 1, 2, 3$. By identifying these bases on each component $\mathscr{V}_i$ of $\mathscr{K}$, we establish a matrix of linear operators $\mathbb{I}_{ij}: \mathscr{V}_j \to \mathscr{V}_i$ satisfying $\mathbb{I}_{ij} d_k = d_k$ for all i, j, and k. Thus, for $i \neq j$ the operator $\mathbb{I}_{ij}$ is an isomorphism identifying the director bases on $\mathscr{V}_i$ and $\mathscr{V}_j$, while for $i = j$ it is just the identity on $\mathscr{V}_i$.

The Euclidean inner product on $\mathscr{V}$ can be used to construct inner products on $\mathscr{Q}$ and $\mathscr{K}$. For any two elements $q_1 = (r_1, d_1)$ and $q_2 = (r_2, d_2)$ of $\mathscr{Q}$, where $d_1 = (d_{11}, d_{21}, d_{31})$ and $d_2 = (d_{12}, d_{22}, d_{32})$, we define

$$q_1 \circ q_2 = r_1 \cdot r_2 + d_1 \circ d_2 \tag{5.1.10}$$

and

$$d_1 \circ d_2 = \sum_{i=1}^{3} d_{i1} \cdot d_{i2}. \tag{5.1.11}$$

Then $\circ$ provides an inner product on $\mathscr{Q}$ and $\mathscr{K}$ with respect to which the component subspaces $\mathscr{V}$, $\mathscr{V}_1$, $\mathscr{V}_2$, and $\mathscr{V}_3$ of $\mathscr{Q}$ are mutually orthogonal.

For mechanical systems, tradition has shown that the Lagrangian can be constructed as the difference

$$L = T - U, \tag{5.1.12}$$

where T and U are the *kinetic* and *potential* energies, respectively. Following the direct approach of Chapter 2, we postulate the quadratic form

$$T = \tfrac{1}{2}\dot{q} \circ \mathbb{E}\dot{q}, \qquad (5.1.13)$$

where $\mathbb{E}$, the *generalized inertia tensor*, is a constant and symmetric element of the space $\mathcal{Q} \otimes \mathcal{Q}$ of second-order tensors on $\mathcal{Q}$. From the perspective of pseudo-rigid bodies, we specialize this assumption further by requiring that $\mathbb{E}$ decompose into a direct sum of operators on $\mathscr{V}$ and $\mathscr{K}$, respectively; namely,

$$\mathbb{E} = m\boldsymbol{I} \oplus \mathbb{E}_{\mathsf{K}}, \qquad (5.1.14)$$

where

$$\mathbb{E}_{\mathsf{K}} = \sum_{i,j=1}^{3} E_{\mathsf{R}}^{ij} \mathbb{I}_i \mathbb{I}_{ij} \mathbb{P}_j. \qquad (5.1.15)$$

m is the total mass, E_{R}^{ij} are the constant components of the Euler tensor E given by $(2.5.19)_1$, and $\mathbb{P}_i \colon \mathscr{K} \to \mathscr{V}_i$ and $\mathbb{I}_i \colon \mathscr{V}_i \to \mathscr{K}$ denote the orthogonal projection and canonical injection maps, respectively. Substituting (5.1.14), (5.1.15), and (5.1.9) into (5.1.13), we obtain for T the expression

$$T = \tfrac{1}{2}m\dot{r} \cdot \dot{r} + \tfrac{1}{2}\dot{d} \circ \mathbb{E}_{\mathsf{K}}\dot{d}$$

$$= \tfrac{1}{2}m\dot{r} \cdot \dot{r} + \tfrac{1}{2}\sum_{i,j=1}^{3} E_{\mathsf{R}}^{ij}\dot{d}_i \cdot \dot{d}_j. \qquad (5.1.16)$$

This is precisely the form of the kinetic energy given by KALONI and DeSILVA [1970] in their work on three-dimensional directed continua.

The potential energy of a mechanical system is the difference between internal and external effects; namely,

$$U = \sigma - u, \qquad (5.1.17)$$

where σ is the stored-energy density

$$\sigma = \sigma(\boldsymbol{d}_1, \boldsymbol{d}_2, \boldsymbol{d}_3) = \sigma(d), \qquad (5.1.18)$$

and u is the load potential

$$u = u(\boldsymbol{r}, \boldsymbol{d}_1, \boldsymbol{d}_2, \boldsymbol{d}_3) = u(q). \qquad (5.1.19)$$

That σ does not depend explicitly on $\boldsymbol{r}$ is a consequence of the principle of material frame indifference.

The Euler–Lagrange equations (5.1.3) reduce in the current notation to the system

$$\frac{d}{dt}\left(\frac{\partial L}{\partial \dot{r}}\right) - \frac{\partial L}{\partial r} = \boldsymbol{0}, \qquad \frac{d}{dt}\left(\frac{\partial L}{\partial \dot{d}_i}\right) - \frac{\partial L}{\partial d_i} = \boldsymbol{0}, \qquad i = 1, 2, 3, \quad (5.1.20)$$

which, by (5.1.12) and (5.1.16)–(5.1.19), becomes

$$\frac{d}{dt}\left(\frac{\partial T}{\partial \dot{r}}\right) - \frac{\partial u}{\partial r} = \boldsymbol{0}, \qquad \frac{d}{dt}\left(\frac{\partial T}{\partial \dot{d}_i}\right) - \frac{\partial u}{\partial d_i} + \frac{\partial \sigma}{\partial d_i} = \boldsymbol{0}, \qquad i = 1, 2, 3. \quad (5.1.21)$$

These Euler–Lagrange equations are the equations of motion of a pseudo-rigid body, the first being that for the center of mass and the others corresponding to the director space $\mathcal{K}$. To demonstrate this explicitly, we first use (5.1.16) to obtain the more explicit system

$$m\ddot{r} = f, \qquad \sum_{j=1}^{3} E_R^{ij}\ddot{d}_j = m^i - \sigma^i, \tag{5.1.22}$$

where we have set

$$f = \frac{\partial u}{\partial r}, \qquad m^i = \frac{\partial u}{\partial d_i}, \qquad \sigma^i = \frac{\partial \sigma}{\partial d_i}. \tag{5.1.23}$$

Clearly, $(5.1.22)_1$ is precisely the balance of linear momentum $(2.3.9)_3$. In order to put $(5.1.22)_2$ into a more familiar form, take its tensor product with d_i and then sum on the index i. This gives us

$$\sum_{i,j=1}^{3} E_R^{ij}\ddot{d}_j \otimes d_i = M - \Sigma, \tag{5.1.24}$$

where M and Σ are defined by

$$M = \sum_{i=1}^{3} m^i \otimes d_i, \qquad \Sigma = \sum_{i=1}^{3} \sigma^i \otimes d_i. \tag{5.1.25}$$

Next we differentiate $(2.5.19)_1$, namely,

$$E = \sum_{i,j=1}^{3} E_R^{ij} d_i \otimes d_j, \tag{5.1.26}$$

and then use $(2.5.18)_2$ to obtain

$$\dot{E} - EL^T = \sum_{i,j=1}^{3} E_R^{ij}\dot{d}_i \otimes d_j. \tag{5.1.27}$$

By the balance law $(2.3.13)_2$ for E this becomes

$$LE = \sum_{i,j=1}^{3} E_R^{ij}\dot{d}_i \otimes d_j. \tag{5.1.28}$$

Differentiating once again and using $(2.5.18)_2$, we obtain

$$(LE)^\cdot - LEL^T = \sum_{i,j=1}^{3} E_R^{ij}\ddot{d}_i \otimes d_j. \tag{5.1.29}$$

Thus, (5.1.24) can be written in the equivalent form

$$(LE)^\cdot - LEL^T = M - \Sigma, \tag{5.1.30}$$

and this is precisely the spatial form $(2.3.13)_4$ of the law of balance of moment of momentum.

It is well known that Euler–Lagrange equations maintain their form under an invertible change of variables $q = Q(\bar{q}, t)$. The natural choice of kinematic variables for a pseudo-rigid body is the pair (r, F) introduced in Chapter 2, and so we transform our Lagrangian formulation now from $q = (r, d)$ to $\bar{q} = (r, F)$. Obviously, this requires only the transform of the director component d of q. To define this change, we recall $(2.5.14)_2$, namely,

$$d_i = F(t)D_i. \tag{5.1.31}$$

Since the D_i are a fixed basis for $\mathscr{V}$, the Lagrangian L now becomes a function on the tangent bundle of $\mathscr{V} \times \mathscr{Gl}(\mathscr{V})$. Therefore, the director component of the Euler–Lagrange equations assumes the alternate form

$$\frac{d}{dt}\left(\frac{\partial L}{\partial \dot{F}}\right) - \frac{\partial L}{\partial F} = \mathbf{0}. \tag{5.1.32}$$

This is a system of differential equations on a configuration manifold $\mathscr{Gl}(\mathscr{V})$ having the structure of a *Lie group*. Its solutions $F = F(t)$ are curves in this manifold, with the corresponding director motions being given by (5.1.31).

Considering the specific case defined by (5.1.16)–(5.1.19), we find that the kinetic energy T takes the form

$$T = \tfrac{1}{2}m\dot{r}\cdot\dot{r} + \tfrac{1}{2}\,\text{tr}\,\dot{F}E_R\dot{F}^{\mathrm{T}}$$
$$= \tfrac{1}{2}m\dot{r}\cdot\dot{r} + \tfrac{1}{2}\dot{F}E_R\cdot\dot{F}, \tag{5.1.33}$$

where the dot in the second term denotes the inner product on $\mathscr{L}(\mathscr{V}, \mathscr{V})$, the space of linear transformations on $\mathscr{V}$, and we have employed (5.1.31) and the identity $\dot{d}_i \cdot \dot{d}_j = \text{tr}\,\dot{d}_i \otimes \dot{d}_j$. The strain energy σ and the load potential u now have the functional forms

$$\sigma = \sigma(F), \qquad u = u(r, F), \tag{5.1.34}$$

respectively, and so from (5.1.32)–(5.1.34) we obtain the equations of motion

$$\dot{H}_R = M_R - \Sigma_R. \tag{5.1.35}$$

H_R is the momentum conjugate to F as given by

$$H_R = \frac{\partial L}{\partial \dot{F}} = \dot{F}E_R, \tag{5.1.36}$$

and we have set

$$M_R = \frac{\partial u}{\partial F}, \qquad \Sigma_R = \frac{\partial \sigma}{\partial F}. \tag{5.1.37}$$

We note that (5.1.35) is precisely $(2.3.9)_4$, the equation of balance of moment of momentum in terms of the reference placement.

Let us set

$$p = \frac{\partial L}{\partial \dot{r}}. \tag{5.1.38}$$

By (5.1.36) and (5.1.38) the canonical momentum corresponding to (r, F) is given by the pair (p, H_R), and so we may form the Hamiltonian $H = H(r, F, p, H_R)$. Then HAMILTON's equations (5.1.6) for a pseudo-rigid body become

$$\dot{r} = \frac{\partial H}{\partial p}, \qquad \dot{p} = -\frac{\partial H}{\partial r} \qquad (5.1.39)$$

for the center of mass, and

$$\dot{F} = \frac{\partial H}{\partial H_R}, \qquad \dot{H}_R = -\frac{\partial H}{\partial F} \qquad (5.1.40)$$

for the director space. For the special case defined by (5.1.12), (5.1.17), and (5.1.33), we find from (5.1.5) that H is given by

$$H = \tfrac{1}{2} p \cdot p m^{-1} + \tfrac{1}{2} H_R \cdot H_R E_R^{-1} + \sigma - u, \qquad (5.1.41)$$

where, consistent with (2.2.22), we have $p = m\dot{r}$. Placing (5.1.41) in (5.1.38) and (5.1.39), we find that HAMILTON's equations become

$$\begin{aligned} \dot{r} = p/m, \qquad & \dot{p} = f, \\ \dot{F} = H_R E_R^{-1}, \qquad & \dot{H}_R = M_R - \Sigma_R, \end{aligned} \qquad (5.1.42)$$

a first-order system clearly consistent with preceding results based on the Euler–Lagrange equations.

The foregoing results are applications of conventional variational methods. A more subtle result arises as follows. Suppose that the motion χ places the body in the reference placement κ at time $t = 0$. Then the specifications $F(0) = I$, $\dot{F}(0) = G_0$, $G_0 \in \mathscr{L}(\mathscr{V}, \mathscr{V})$, are initial conditions for the equations of motion

$$\ddot{F} E_R = M_R - \Sigma_R. \qquad (5.1.43)$$

The set of generalized velocities G_0 makes up the tangent space to $\mathscr{Gl}(\mathscr{V})$ at the identity I. This linear space can be given the structure of a Lie algebra, often called the *standard Lie algebra* $gl(\mathscr{V})$ of $\mathscr{Gl}(\mathscr{V})$. The Lie algebra at an arbitrary point F of $\mathscr{Gl}(\mathscr{V})$ is identified with $gl(\mathscr{V})$ by means of the isomorphism identifying G_0 with $\dot{F}F^{-1}$. But the latter tensor is precisely L, the deforming tensor. While the pair $(F, \dot{F})$ provides appropriate variables with which to obtain the equations of motion relative to κ, the preceding fact suggests that the pair (F, L) might provide variables more appropriate to obtaining the equations of motion relative to the current placement.

Motivated by these remarks, we reexpress the Lagrangian in terms of a pair (F, L) for which there is an invertible transformation between $\dot{F}$ and L. For the moment, consider a general linear transformation

$$L = \mathbb{A}(F)\dot{F}, \qquad (5.1.44)$$

where $\mathbb{A} \in \mathscr{Gl}(\mathscr{L}(\mathscr{V}, \mathscr{V}))$ is a fourth-order tensor-valued function of F. It is important to note that (5.1.44) is generally not compatible with an allowable change of variables of the form $q = Q(\bar{q}, t)$, and so we may not expect the Euler–Lagrange equations to remain invariant. This situation is not unfamiliar to Lagrangian dynamics. The variable L is commonly called a *quasivelocity*, and the Euler–Lagrange equations reformulated in terms of F and L are said to be given in terms of *quasivariables*. We turn now to the derivation of these equations when (5.1.44) applies.

Let L^F denote the Lagrangian appearing in (5.1.32) and which expresses L as a function of F and $\dot{F}$. Define a new Lagrangian L^L by setting

$$L^L(F, L) = L^F(F, \mathbb{A}^{-1}(F)L). \tag{5.1.45}$$

By a judicious use of differential calculus, the chain rule, and appropriate manipulations of tensors, we find that the Euler–Lagrange equations expressed in terms of L^L take the form

$$\frac{d}{dt}\left(\frac{\partial L^L}{\partial L}\right) - \mathbb{A}^{-\mathsf{T}}\frac{\partial L^L}{\partial F} = -2\mathbb{A}^{-\mathsf{T}}\operatorname{sk}\left(\frac{\partial(\mathbb{A}^{\mathsf{T}}H)}{\partial F}\right)_H(\mathbb{A}^{-1}L), \tag{5.1.46}$$

where

$$H = \frac{\partial L^L}{\partial L} \tag{5.1.47}$$

is the *conjugate quasimomentum*. The subscript H on the derivative in (5.1.46) means that H is regarded as constant during the differentiation with respect to F. The transpose $\mathbb{A}^{\mathsf{T}}$ of a fourth-order tensor is defined in the usual way, namely, by requiring that $\mathbb{A}C \cdot D = C \cdot \mathbb{A}^{\mathsf{T}}D$ for all C, $D \in \mathscr{L}(\mathscr{V}, \mathscr{V})$. Thus, for example, $(\mathbb{A}\mathbb{B})^{\mathsf{T}} = \mathbb{B}^{\mathsf{T}}\mathbb{A}^{\mathsf{T}}$.

Consider now the special case in which (5.1.44) is given by

$$L = \dot{F}F^{-1}. \tag{5.1.48}$$

In this case, $\mathbb{A}$ can be written in terms of notation introduced by DEL PIERO [1979] as the fourth-order tensor

$$\mathbb{A} = I \boxtimes F^{-\mathsf{T}}. \tag{5.1.49}$$

DEL PIERO has established the following rules for such tensors, where we can interpret the first as a definition of the product $\boxtimes$:

$$\begin{aligned}
(A \boxtimes B)C &= ACB^{\mathsf{T}}, \\
(A \boxtimes B)^{\mathsf{T}} &= A^{\mathsf{T}} \boxtimes B^{\mathsf{T}}, \\
(A \boxtimes B)^{-1} &= A^{-1} \boxtimes B^{-1}, \\
(A \boxtimes B)\mathbb{T} &= \mathbb{T}(B \boxtimes A),
\end{aligned} \tag{5.1.50}$$

where A, B, C are any second-order tensors and $\mathbb{T}$ is the fourth-order transposition tensor defined by $\mathbb{T} A = A^T$. A simple computation and these properties give us the derivatives

$$\frac{\partial F^{-1}}{\partial F} = -(F^{-1} \boxtimes F^{-T}), \quad \frac{\partial F^{-T}}{\partial F} = \mathbb{T}\frac{\partial F^{-1}}{\partial F} = -(F^{-T} \boxtimes F^{-1})\mathbb{T}, \quad (5.1.51)$$

and therefore the right-hand side of (5.1.46) reduces with the choice (5.1.49) to

$$-2\mathbb{A}^{-T} \operatorname{sk}\left(\frac{\partial(\mathbb{A}^T H)}{\partial F}\right)_H (\mathbb{A}^{-1} L) = HL^T - L^T H. \quad (5.1.52)$$

Thus, the Euler–Lagrange equations (5.1.46) simplify in this case to

$$\frac{d}{dt}\left(\frac{\partial L^L}{\partial L}\right) - \frac{\partial L^L}{\partial F}F^T = \frac{\partial L^L}{\partial L}L^T - L^T\frac{\partial L^L}{\partial L}. \quad (5.1.53)$$

By (5.1.48) the kinetic energy (5.1.33) takes the form

$$T = \tfrac{1}{2}m\dot{r}\cdot\dot{r} + \tfrac{1}{2}L\cdot\mathbb{E}L$$

$$= \tfrac{1}{2}m\dot{r}\cdot\dot{r} + \tfrac{1}{2}(LE)\cdot L, \quad (5.1.54)$$

where the fourth-order inertia tensor $\mathbb{E}$ is given by

$$\mathbb{E} = I \boxtimes E. \quad (5.1.55)$$

From $(5.1.54)_2$, (5.1.47), and (2.2.16), namely,

$$E = FE_R F^T, \quad (5.1.56)$$

we find that

$$\frac{\partial T}{\partial F} = L^T HF^{-T},$$

$$\frac{\partial T}{\partial L} = \frac{\partial L^L}{\partial L} = H = LE. \quad (5.1.57)$$

Finally, by substituting (5.1.57) and (5.1.34) into (5.1.53), we obtain

$$(LE)^{\cdot} - LEL^T = M - \Sigma, \quad (5.1.58)$$

where

$$M = \frac{\partial u}{\partial F}F^T, \quad \Sigma = \frac{\partial \sigma}{\partial F}F^T. \quad (5.1.59)$$

By virtue of (5.1.37) we see that the identifications (5.1.59) are consistent with the transformation equations (2.3.8) relating the reference and current representations of M and Σ. Of course, (5.1.58) are the equations of motion relative to the current placement (cf. $(2.3.13)_4$).

(ii) Symmetry and Conserved Quantities

The theory of pseudo-rigid bodies, like other mechanical theories, possesses special symmetries common to most natural systems. The fact that different observers should see essentially the same deformation, "essentially" meaning that observations should transform according to the usual rules of transformation in a change of frame, leads by use of the principle of material frame indifference to a symmetry of the equations of motion. More precisely, *any superposed rigid motion applied to one solution leads to another solution.* Similarly, for materials that are isotropic, *a change of frame in the reference placement of a body transforms solutions of the equations of balance into other solutions.* Since the advent of modern continuum mechanics, definite rules have been devised for building specific symmetries into a broad class of mechanical theories. These are phrased[1] principally in terms of restrictions on the response function of the material involved. Unfortunately, general methods for exploiting the invariance of a system of equations are much more elusive. One exception appears in the variational calculus. The well-known theorems of EMMY NOETHER [1918] provide a means of using invariance properties directly through the construction of conserved quantities, or in more general settings through the construction of conservation laws (cf. MARSDEN and HUGHES [1983]). Often these conserved quantities have important physical interpretations and are fundamental ingredients in the analysis of the associated dynamic equations.

Our objective here is to develop Noether-like theorems for the equations of motion of a pseudo-rigid body. However, the technique we use is unconventional. We wish to show that symmetries can be addressed more directly through the *calculus of variations on a group.* By introducing the elements of a group as variables in a variational problem, we find that Euler–Lagrange equations arise that are more naturally suited to the problem. These are important in their own right, but if additionally the variational problem is invariant under this group, then out of these equations we obtain immediately the conserved quantities first proposed by NOETHER.

The calculus of variations on a group can be illustrated most easily in terms of the deformation F. Conventionally we would regard the Lagrangian L as a function of F and $\dot{F}$ (suppressing for the moment r and $\dot{r}$), and the corresponding Euler–Lagrange equations would take the form (5.1.32). However, the use of $\dot{F}$ suggests that F is an element of a linear space, while it is more natural to regard F as an element of the general linear group $\mathscr{Gl}(\mathscr{V})$. We convert to a formulation on this group

---

[1] As HEALEY [1988] has noted, the invariance of the equations of motion in a mechanical theory, while certainly requiring special transformation rules for response functions, is a more global property of a system. Special symmetry is also required of the shape of the body, loads applied to it, etc.

in two steps. First, we eliminate the rate variable $\dot{F}$ in favor of the associated element in the Lie algebra of the group. As indicated following (5.1.43), this means that we use $\dot{F}F^{-1}$, or the deformation tensor L, instead of $\dot{F}$. Equivalently, we work in terms of a particular quasivelocity. This gives rise to the new Lagrangian L^{L} defined generally by (5.1.45).

The Euler–Lagrange equations based on L^{L} can be obtained by the general method of quasivelocities outlined in Section 5(i). The result is (5.1.53). However, a different route to this equation is more illuminating, and this is the second change we make to the conventional approach. When working in a linear space, a *variation* of $F(t)$ is a one-parameter family of deformations of the form $F(t) + \varepsilon G(t)$. When working on a group, however, it is more natural to use one-parameter subgroups. We define a variation $F^{\varepsilon}(t)$ of $F(t)$ to be a function of the form

$$F^{\varepsilon}(t) = G(\varepsilon, t)F(t), \qquad (5.2.1)$$

where G defines any time-dependent one-parameter subgroup of $\mathscr{Gl}(\mathscr{V})$. This means that G is a solution of the differential equation

$$\frac{d}{d\varepsilon}G(\varepsilon, t) = A(t)G(\varepsilon, t), \qquad G(0, t) = I, \qquad (5.2.2)$$

where $A(t)$ is an arbitrary element of the Lie algebra for each t. Equivalently,

$$G(\varepsilon, t) = \exp \varepsilon A(t). \qquad (5.2.3)$$

The corresponding variation of the deforming tensor L is

$$
\begin{aligned}
L^{\varepsilon} &= \dot{F}^{\varepsilon}(F^{\varepsilon})^{-1} \\
&= (\dot{G}F + G\dot{F})F^{-1}G^{-1} \\
&= \dot{G}G^{-1} + GLG^{-1}, \qquad (5.2.4)
\end{aligned}
$$

where the over dot on G denotes a partial derivative with respect to t. If δ denotes a partial derivative with respect to ε evaluated at $\varepsilon = 0$, we see from (5.2.1), (5.2.2), and (5.2.4) that

$$\delta F^{\varepsilon} = AF, \qquad \delta L^{\varepsilon} = \dot{A} + AL - LA. \qquad (5.2.5)$$

By the chain rule we obtain

$$
\begin{aligned}
&\delta L^{L}(F^{\varepsilon}, L^{\varepsilon}) \\
&= \frac{\partial L^{L}}{\partial F} \cdot \delta F^{\varepsilon} + \frac{\partial L^{L}}{\partial L} \cdot \delta L^{\varepsilon} \\
&= \frac{\partial L^{L}}{\partial F} \cdot AF + \frac{\partial L^{L}}{\partial L} \cdot (\dot{A} + AL - LA) \\
&= \frac{d}{dt}\left(\frac{\partial L^{L}}{\partial L} \cdot A\right) - \left(\frac{d}{dt}\left(\frac{\partial L^{L}}{\partial L}\right) - \frac{\partial L^{L}}{\partial F}F^{\mathrm{T}} - \frac{\partial L^{L}}{\partial L}L^{\mathrm{T}} + L^{\mathrm{T}}\frac{\partial L^{L}}{\partial L}\right) \cdot A. \quad (5.2.6)
\end{aligned}
$$

As an element of the Lie algebra of $\mathscr{G}\ell(\mathscr{V})$, the matrix $A(t)$ is arbitrary. We conclude that the conventional variational equation

$$0 = \delta \int_{\mathscr{I}} L^L(F^\varepsilon, L^\varepsilon), \qquad \mathscr{I} = (t_1, t_2), \tag{5.2.7}$$

reduces to the system of Euler–Lagrange equations (5.1.53) and the natural boundary condition that $\partial L^L/\partial L$ vanishes at t_1 and t_2.

Recalling the equivalence between (5.1.53) and (5.1.58), we have shown that *the equations of motion for pseudo-rigid bodies, phrased in terms of the current placement, are Euler–Lagrange equations on the general linear group $\mathscr{G}\ell(\mathscr{V})$*. It should be noted that (5.1.53) is a system of first-order differential equations in L and F. In order to use it in applications, we must also use the equation

$$\dot{F} = LF \tag{5.2.8}$$

that follows from the definition of L.

A variant of the preceding analysis leads to Euler–Lagrange equations of a slightly different structure. Rather than using $L = \dot{F}F^{-1}$ as a general element of the Lie algebra of $\mathscr{G}\ell(\mathscr{V})$, we work in terms of the tensor $K = F^{-1}\dot{F}$. In this case, (5.2.8) is replaced by

$$\dot{F} = FK. \tag{5.2.9}$$

The associated variation K^ε is given by

$$\begin{aligned} K^\varepsilon &= (F^\varepsilon)^{-1}\dot{F}^\varepsilon \\ &= F^{-1}G^{-1}(\dot{G}F + G\dot{F}) \\ &= K + F^{-1}G^{-1}\dot{G}F, \end{aligned} \tag{5.2.10}$$

and therefore

$$\delta K^\varepsilon = F^{-1}\dot{A}F. \tag{5.2.11}$$

Let L^K denote the Lagrangian expressed as a function of F and K. Then

$$\begin{aligned} &\delta L^K(F^\varepsilon, K^\varepsilon) \\ &= \frac{\partial L^K}{\partial F} \cdot \delta F^\varepsilon + \frac{\partial L^K}{\partial K} \cdot \delta K^\varepsilon \\ &= \frac{\partial L^K}{\partial F} \cdot AF + \frac{\partial L^K}{\partial K} \cdot F^{-1}\dot{A}F \\ &= \frac{\partial L^K}{\partial F}F^{\mathrm{T}} \cdot A + F^{-\mathrm{T}}\frac{\partial L^K}{\partial K}F^{\mathrm{T}} \cdot \dot{A} \\ &= \frac{d}{dt}\left(F^{-\mathrm{T}}\frac{\partial L^K}{\partial K}F^{\mathrm{T}} \cdot A\right) - \left(\frac{d}{dt}\left(F^{-\mathrm{T}}\frac{\partial L^K}{\partial K}F^{\mathrm{T}}\right) - \frac{\partial L^K}{\partial F}F^{\mathrm{T}}\right) \cdot A. \end{aligned} \tag{5.2.12}$$

As in the previous calculation, the second term on the right-hand side must vanish. This leads to the system of Euler–Lagrange equations

$$\frac{d}{dt}\left(F^{-\text{T}}\frac{\partial L^{\text{K}}}{\partial K}F^{\text{T}}\right) = \frac{\partial L^{\text{K}}}{\partial F}F^{\text{T}}. \tag{5.2.13}$$

As we shall see later, it is the fact that the left-hand side here is a total derivative with respect to time that leads to conservation principles for invariant problems.

The foregoing analyses exploit the fact that a variational problem is naturally phrased on a group, but they say nothing about the implications of invariance. For this it is more useful to focus on the orthogonal group $\mathcal{O}(\mathcal{V})$ rather than the full linear group. Consider the polar decomposition of the deformation,

$$F = RU. \tag{5.2.14}$$

We view this as a change of variables in which F is replaced by (1) an element R of $\mathcal{O}(\mathcal{V})$ and (2) an element U of the space $\mathcal{S}^{+}(\mathcal{V})$ of positive-definite symmetric tensors. R is treated as a group variable with corresponding quasivelocity $\Omega = R^{\text{T}}\dot{R}$. Naturally, Ω lies in the space $\mathcal{A}(\mathcal{V})$ of skew tensors on $\mathcal{V}$, this being the Lie algebra of $\mathcal{O}(\mathcal{V})$. In contrast, we view U as an element of a linear space and so use the generalized velocity $\dot{U}$ directly. A typical variation of F then has the form

$$F^{\varepsilon}(t) = Q(\varepsilon, t)R(t)(U(t) + \varepsilon V(t)), \tag{5.2.15}$$

where

$$Q(\varepsilon, t) = \exp \varepsilon A(t), \tag{5.2.16}$$

and V and A are arbitrary time-dependent symmetric and skew tensors, respectively. The advantage of this type of variation is the fact that for ε sufficiently close to 0 its polar decomposition is clear: if

$$F^{\varepsilon} = R^{\varepsilon}U^{\varepsilon}, \tag{5.2.17}$$

then

$$R^{\varepsilon}(t) = Q(\varepsilon, t)R(t), \qquad U^{\varepsilon}(t) = U(t) + \varepsilon V(t). \tag{5.2.18}$$

Let the Lagrangian be expressed as a function L^{U} of U, $\dot{U}$, R, and Ω. The corresponding Euler–Lagrange equations can be deduced easily from our preceding calculations. Since $R^{\text{T}}\dot{R}$ is the analogue for $\mathcal{O}(\mathcal{V})$ of $F^{-1}\dot{F}$ for $\mathcal{Gl}(\mathcal{V})$, we find from (5.2.12) and the usual calculations in the variational calculus that

$$\delta L^{\text{U}}(U^{\varepsilon}, \dot{U}^{\varepsilon}, R^{\varepsilon}, \Omega^{\varepsilon}) = \frac{d}{dt}\left(R\frac{\partial L^{\text{U}}}{\partial \Omega}R^{\text{T}}\cdot A + \frac{\partial L^{\text{U}}}{\partial \dot{U}}\cdot V\right)$$

$$-\left(\frac{d}{dt}\left(R\frac{\partial L^{\text{U}}}{\partial \Omega}R^{\text{T}}\right) - \frac{\partial L^{\text{U}}}{\partial R}R^{\text{T}}\right)\cdot A$$

$$-\left(\frac{d}{dt}\left(\frac{\partial L^{\text{U}}}{\partial \dot{U}}\right) - \frac{\partial L^{\text{U}}}{\partial U}\right)\cdot V. \tag{5.2.19}$$

Since A and V are arbitrary skew and symmetric tensors, respectively, we obtain the Euler–Lagrange equations in the form

$$\text{sk}\left(\frac{d}{dt}\left(R\frac{\partial L^{U}}{\partial \Omega}R^{\mathrm{T}}\right) - \frac{\partial L^{U}}{\partial R}R^{\mathrm{T}}\right) = 0,$$

$$\text{sym}\left(\frac{d}{dt}\left(\frac{\partial L^{U}}{\partial \dot{U}}\right) - \frac{\partial L^{U}}{\partial U}\right) = 0.$$

(5.2.20)

Naturally, we must append to these the analogue of (5.2.9) for $\mathcal{O}(\mathcal{V})$, namely,

$$\dot{R} = R\Omega.$$

(5.2.21)

The partial derivatives of L^{U} appearing in (5.2.20) require special interpretation. From an abstract perspective, we view $\partial L^{U}/\partial \Omega$ as a linear functional on the space of skew tensors $\mathcal{A}(\mathcal{V})$. Its value $(\partial L^{U}/\partial \Omega)\cdot \Lambda$ at another skew tensor Λ is a linear approximation of $L^{U}(U, \dot{U}, R, \Omega + \Lambda)$ for Λ small. In this approach to derivatives, we take $\partial L^{U}/\partial \Omega$ to be a tensor in the space on which it operates; that is, $\partial L^{U}/\partial \Omega$ is assumed to be a skew tensor. In a similar way, we interpret $\partial L^{U}/\partial U$ and $\partial L^{U}/\partial \dot{U}$ as symmetric tensors. From a computational perspective, we conventionally form the above derivatives from component equivalents $\partial L^{U}/\partial \Omega_{ij}$, $\partial L^{U}/\partial U_{ij}$, and $\partial L^{U}/\partial \dot{U}_{ij}$ relative to a basis where, for the purpose of forming the derivatives, the nine components of each of Ω, U, and $\dot{U}$ are regarded as independent variables. In order to link these two perspectives we adopt the conventions

$$\left[\frac{\partial L^{U}}{\partial \Omega}\right] = \left(\frac{1}{2}\frac{\partial L^{U}}{\partial \Omega_{ij}} - \frac{1}{2}\frac{\partial L^{U}}{\partial \Omega_{ji}}\right),$$

$$\left[\frac{\partial L^{U}}{\partial U}\right] = \left(\frac{1}{2}\frac{\partial L^{U}}{\partial U_{ij}} + \frac{1}{2}\frac{\partial L^{U}}{\partial U_{ji}}\right),$$

(5.2.22)

with a similar formula for $\partial L^{U}/\partial \dot{U}$. In these conventions, $[T]$ denotes a matrix representation of T relative to a fixed orthogonal basis $D_i \otimes D_j$ of $\mathcal{V} \otimes \mathcal{V}$ (cf. Section 4(i)). In either case, (5.2.20) can now be written in the simpler form

$$\frac{d}{dt}\left(R\frac{\partial L^{U}}{\partial \Omega}R^{\mathrm{T}}\right) = \text{sk}\left(\frac{\partial L^{U}}{\partial R}R^{\mathrm{T}}\right),$$

$$\frac{d}{dt}\left(\frac{\partial L^{U}}{\partial \dot{U}}\right) = \frac{\partial L^{U}}{\partial U},$$

(5.2.23)

where the first is an equation on $\mathcal{A}(\mathcal{V})$ and the second applies only on $\mathcal{S}(\mathcal{V})$.

If the triple product in (5.2.23) is differentiated explicitly with respect to t, and time derivatives of R are eliminated by use of (5.2.21), we obtain the following alternate equation:

$$\frac{d}{dt}\left(\frac{\partial L^U}{\partial \Omega}\right) = \frac{\partial L^U}{\partial \Omega}\Omega - \Omega\frac{\partial L^U}{\partial \Omega} + R^T \text{ sk}\left(\frac{\partial L^U}{\partial R}R^T\right)R. \qquad (5.2.24)$$

Consider now the implications of the principle of material frame indifference. According to $(2.7.13)_3$, the stored-energy function σ can be viewed as a function σ^u of U alone. Therefore, if we suppress once again the variables r and $\dot{r}$, (5.1.12), (5.1.17), and (5.1.33) give us the Lagrangian

$$\begin{aligned} L^U &= \tfrac{1}{2}\dot{F}E_R\cdot\dot{F} - \sigma(F) + u(F) \\ &= \tfrac{1}{2}(\dot{R}U + R\dot{U})E_R\cdot(\dot{R}U + R\dot{U}) - \sigma^u(U) + u(RU) \\ &= \tfrac{1}{2}R(\dot{U} + \Omega U)E_R\cdot R(\dot{U} + \Omega U) - \sigma^u(U) + u(RU) \\ &= \tfrac{1}{2}(\dot{U} + \Omega U)E_R\cdot(\dot{U} + \Omega U) - \sigma^u(U) + u(RU). \qquad (5.2.25) \end{aligned}$$

Note that if u is absent, or if it possesses the same type of invariance as σ, then R is absent. More generally, we note that the polar decomposition of QF is $(QR)U$. Moreover, $(QR)^T(QR)^{\cdot} = R^TQ^TQ\dot{R} = \Omega$ provided Q is independent of time. This shows that a superposed rigid-body displacement is defined by the transformation $U\mapsto U$, $\dot{U}\mapsto\dot{U}$, $R\mapsto QR$, $\Omega\mapsto\Omega$, and therefore restricted material frame indifference[2] of the stored energy σ and the load potential u imply the identity

$$L^U(U, \dot{U}, R, \Omega) = L^U(U, \dot{U}, QR, \Omega). \qquad (5.2.26)$$

This is a formal expression of the fact that the Lagrangian L^U is *invariant with respect to rigid-body displacements*. By choosing $Q = R^T$, we conclude that in any system with this invariance the Lagrangian is independent of R.

The implications of invariance under rigid-body displacements can be seen immediately from $(5.2.23)_1$; namely,

$$R\frac{\partial L^U}{\partial \Omega}R^T = C, \qquad (5.2.27)$$

where C is a constant skew tensor. This is equivalent to the three conserved quantities that NOETHER's results deliver corresponding to invariance under the three-parameter group $\mathcal{O}(\mathcal{V})$. Unlike conventional approaches, however, we have obtained them by introducing the elements of the group explicitly as variables in the variational problem.

Physically, (5.2.27) states that the angular momentum of the pseudo-rigid body is constant. Indeed, (5.1.36) and (2.2.21) show that

$$H = \frac{\partial L^F}{\partial \dot{F}}F^T. \qquad (5.2.28)$$

Since $L^U(U, \dot{U}, R, \Omega) = L^F(RU, R(\dot{U} + \Omega U))$, a differentiation and use of the convention $(5.2.22)_1$ imply that

---

[2] Cf. Section 3(iv).

$$\frac{\partial L^{\mathrm{U}}}{\partial \boldsymbol{\Omega}} = \mathrm{sk}\left(\boldsymbol{R}^{\mathrm{T}} \frac{\partial L^{\mathrm{F}}}{\partial \dot{\boldsymbol{F}}} \boldsymbol{U} \right), \tag{5.2.29}$$

and therefore,

$$\boldsymbol{R} \frac{\partial L^{\mathrm{U}}}{\partial \boldsymbol{\Omega}} \boldsymbol{R}^{\mathrm{T}} = \mathrm{sk}\left(\frac{\partial L^{\mathrm{F}}}{\partial \dot{\boldsymbol{F}}} \boldsymbol{U} \boldsymbol{R}^{\mathrm{T}} \right) = \mathrm{sk}\ \boldsymbol{H}. \tag{5.2.30}$$

By (2.5.9) this is the angular momentum A of the pseudo-rigid body. Moreover, (5.2.25) shows that

$$\frac{\partial L^{\mathrm{U}}}{\partial \boldsymbol{R}} \boldsymbol{R}^{\mathrm{T}} = \frac{\partial}{\partial \boldsymbol{R}}(u(\boldsymbol{R}\boldsymbol{U}))\boldsymbol{R}^{\mathrm{T}}$$

$$= \frac{\partial u}{\partial \boldsymbol{F}} \boldsymbol{U} \boldsymbol{R}^{\mathrm{T}}$$

$$= \frac{\partial u}{\partial \boldsymbol{F}} \boldsymbol{F}^{\mathrm{T}}, \tag{5.2.31}$$

and therefore, $(5.1.59)_1$ gives us

$$\mathrm{sk}\left(\frac{\partial L^{\mathrm{U}}}{\partial \boldsymbol{R}} \boldsymbol{R}^{\mathrm{T}} \right) = \mathrm{sk}\ \boldsymbol{M}. \tag{5.2.32}$$

The results (5.2.30) and (5.2.32) show that the Euler–Lagrange equation $(5.2.23)_1$ is precisely the balance of angular momentum (2.5.10). Moreover, (5.2.27) asserts that *the angular momentum of a pseudo-rigid body is constant in any torque-free motion.*

For certain calculations, the alternate equation (5.2.24) is more useful. If ω denotes the axial vector corresponding to the skew tensor $\boldsymbol{\Omega}$, we may view the Lagrangian instead as a function of $\boldsymbol{U}$, $\dot{\boldsymbol{U}}$, and ω. For simplicity, we denote this function by L. In this case, the axial vector equation corresponding to (5.2.24) is given by

$$\frac{d}{dt}\left(\frac{\partial L}{\partial \omega} \right) = -\omega \times \frac{\partial L}{\partial \omega} + \tau_{\mathrm{R}}, \qquad \tau_{\mathrm{R}} = \boldsymbol{R}^{\mathrm{T}} \tau, \tag{5.2.33}$$

where τ is the net torque given by the axial vector corresponding to $2\ \mathrm{sk}\ \boldsymbol{M}$. This equation is the analogue for pseudo-rigid bodies of the system (2.5.36) representing EULER'S equations for a rigid body. However, the analogy becomes precise only when we draw a connection between the axial vectors ω and $\boldsymbol{w}$. Recall that $\boldsymbol{w}$ is the axial vector of the angular velocity tensor $\boldsymbol{W} = \dot{\boldsymbol{R}}\boldsymbol{R}^{\mathrm{T}}$. Since $\boldsymbol{\Omega} = \boldsymbol{R}^{\mathrm{T}}\dot{\boldsymbol{R}} = \boldsymbol{R}^{\mathrm{T}}\boldsymbol{W}\boldsymbol{R}$, we see that[3]

---

[3] As pointed out in Section 2(vi), there is a dualism in mechanics that can be stated, in the present context, as follows. The *active* view regards $\boldsymbol{W}$ and $\boldsymbol{w}$ as angular velocities of the body relative to a fixed frame of reference. The *passive* view interprets the motion of the body through a change of frame. Evidently, the angular velocities of the moving frame would be $-\boldsymbol{W}$ and $-\boldsymbol{w}$, respectively. Here our viewpoint is always the active one, and the dualism that arises stems from our use of a reference placement. The angular velocities $\boldsymbol{\Omega}$ and ω, the "pull-backs" of $\boldsymbol{W}$ and $\boldsymbol{w}$, are required so that an observer in κ sees the same velocities relative to the body as those observed in the current placement.

$\omega = R^{\mathsf{T}} w$. In Section 2(v) we interpreted a basis D_i in the reference placement of the body as a set of axes fixed in space, and in terms of it we introduced body axes as those given by a basis d_i defined by $(2.5.14)_2$. These basis vectors rotate, stretch, and shear with the body in its motion. It is more convenient here to introduce *corotating body axes*, defined by the basis

$$\hat{d_i} = R(t)D_i, \qquad i = 1, 2, 3. \tag{5.2.34}$$

These vectors rotate with the body but ignore any stretching and shearing that may be occurring. If the basis D_i is orthonormal, the basis $\hat{d_i}$ is also orthonormal. Let $\hat{w}_i$ denote the components of the angular velocity w relative to corotating body axes:

$$\hat{w}_i = w \cdot \hat{d_i}. \tag{5.2.35}$$

Then

$$\omega \cdot D_i = R^{\mathsf{T}} w \cdot R^{\mathsf{T}} \hat{d_i} = w \cdot \hat{d_i} = \hat{w}_i. \tag{5.2.36}$$

That is, the components of ω relative to the axes D_i fixed in space are exactly the components of the angular velocity relative to corotating axes. It is well known in rigid-body mechanics that EULER's equations (2.5.36) are most conveniently expressed in terms of these components of angular velocity. Indeed, the codeforming derivative that appears in (2.5.36) has been introduced precisely so that we may work relative to corotating axes. Equation (5.2.33) attains this same end within the framework of the calculus of variations.

The preceding analogy can be strengthened further by restricting attention completely to rigid-body mechanics. When the net torque is zero, the Lagrangian is formed purely from kinetic energy and has the form

$$L = \tfrac{1}{2} w \cdot Jw. \tag{5.2.37}$$

Since $\omega = R^{\mathsf{T}} w$, this becomes

$$L = \tfrac{1}{2} R\omega \cdot JR\omega$$
$$= \tfrac{1}{2} \omega \cdot J_{\mathrm{R}} \omega, \tag{5.2.38}$$

where, in analogy with (2.2.16), we define the *referential inertia tensor*

$$J_{\mathrm{R}} = R^{\mathsf{T}} JR. \tag{5.2.39}$$

Relative to axes D_i fixed in space, the components of J_{R} are constant; that is, they are precisely the components of J relative to corotating axes. Thus, (5.2.33) reduces in this case to

$$J_{\mathrm{R}} \dot{\omega} = -\omega \times J_{\mathrm{R}} \omega, \tag{5.2.40}$$

and relative to fixed axes these are exactly EULER's equations.

We have focused so far on the group of rigid-body displacements because of its basic role in the general theory for any pseudo-rigid body. However, additional invariance can arise if the body is isotropic and the Euler tensor exhibits an axis of symmetry. In order to analyze symmetries of this type, while retaining the results given above for $\mathcal{O}(\mathcal{V})$, we extend the decomposition (5.2.14) as follows. Consider a *double polar decomposition of F* given by

$$F = R_L D R_R, \qquad (5.2.41)$$

in which (1) D has a positive-definite diagonal matrix representation relative to the basis D_i, and (2) R_L and R_R are orthogonal transformations that operate on D on the left and on the right, respectively. In place of $\dot{R}_L$ we use $\Omega_L = R_L^T \dot{R}_L$ and in place of $\dot{R}_R$ we use $\Omega_R = \dot{R}_R R_R^T$. D is viewed as an element of a linear space, and so we use its standard generalized velocity $\dot{D}$. In terms of these conventions, $\dot{F}$ is given by

$$\dot{F} = \dot{R}_L D R_R + R_L \dot{D} R_R + R_L D \dot{R}_R$$

$$= R_L (\dot{D} + \Omega_L D + D \Omega_R) R_R, \qquad (5.2.42)$$

and then we form a new Lagrangian L^D by the definition

$$L^D(D, \dot{D}, R_L, \Omega_L, R_R, \Omega_R) = L^F(R_L D R_R, R_L(\dot{D} + \Omega_L D + D \Omega_R) R_R). \quad (5.2.43)$$

Consider variations of F of the form

$$F^\varepsilon(t) = Q_L(\varepsilon, t) R_L(t) (D(t) + \varepsilon \Delta(t)) R_R(t) Q_R(\varepsilon, t), \qquad (5.2.44)$$

in which Q_L and Q_R are given by one-parameter subgroups of $\mathcal{O}(\mathcal{V})$. Thus,

$$Q_L(\varepsilon, t) = \exp \varepsilon A_L(t), \qquad Q_R(\varepsilon, t) = \exp \varepsilon A_R(t), \qquad (5.2.45)$$

where A_L and A_R are any time-dependent skew tensors. Equivalently, F^ε has the double polar decomposition $R_L^\varepsilon D^\varepsilon R_R^\varepsilon$, where

$$R_L^\varepsilon = Q_L R_L, \qquad D^\varepsilon = D + \varepsilon \Delta, \qquad R_R^\varepsilon = R_R Q_R. \qquad (5.2.46)$$

It follows that the variations in Ω_L and Ω_R are

$$\Omega_L^\varepsilon = R_L^{\varepsilon T} \dot{R}_L^\varepsilon = R_L^T Q_L^T (\dot{Q}_L R_L + Q_L \dot{R}_L) = \Omega_L + R_L^T Q_L^T \dot{Q}_L R_L,$$

$$\Omega_R^\varepsilon = \dot{R}_R^\varepsilon R_R^{\varepsilon T} = (\dot{R}_R Q_R + R_R \dot{Q}_R) Q_R^T R_R^T = \Omega_R + R_R \dot{Q}_R Q_R^T R_R^T, \qquad (5.2.47)$$

and therefore,

$$\delta R_L^\varepsilon = A_L R_L, \qquad \delta D^\varepsilon = \Delta, \qquad \delta R_R^\varepsilon = R_R A_R,$$

$$\delta \Omega_L^\varepsilon = R_L^T \dot{A}_L R_L, \qquad \delta \Omega_R^\varepsilon = R_R \dot{A}_R R_R^T. \qquad (5.2.48)$$

Note the analogy between $(5.2.48)_4$ and $(5.2.11)$. By the chain rule, the variation in the Lagrangian is given, in analogy with $(5.2.19)$, by

$$\delta L^D(D^\epsilon, \dot{D}^\epsilon, R_L^\epsilon, \Omega_L^\epsilon, R_R^\epsilon, \Omega_R^\epsilon)$$

$$= \frac{d}{dt}\left(R_L \frac{\partial L^D}{\partial \Omega_L} R_L^T \cdot A_L + \frac{\partial L^D}{\partial \dot{D}} \cdot \Delta + R_R^T \frac{\partial L^D}{\partial \Omega_R} R_R \cdot A_R \right)$$

$$- \left(\frac{d}{dt}\left(R_L \frac{\partial L^D}{\partial \Omega_L} R_L^T \right) - \frac{\partial L^D}{\partial R_L} R_L^T \right) \cdot A_L - \left(\frac{d}{dt}\left(\frac{\partial L^D}{\partial \dot{D}} \right) - \frac{\partial L^D}{\partial D} \right) \cdot \Delta$$

$$- \left(\frac{d}{dt}\left(R_R^T \frac{\partial L^D}{\partial \Omega_R} R_R \right) - R_R^T \frac{\partial L^D}{\partial R_R} \right) \cdot A_R. \tag{5.2.49}$$

Let us extend the conventions (5.2.22) by setting

$$\left[\frac{\partial L^D}{\partial D} \right] = \text{diag}\left(\frac{\partial L^D}{\partial D_{11}}, \frac{\partial L^D}{\partial D_{22}}, \frac{\partial L^D}{\partial D_{33}} \right) \tag{5.2.50}$$

relative to a prescribed basis $D_i \otimes D_j$, with a similar formula for $\partial L^D/\partial \dot{D}$. Then, since A_L and A_R are arbitrary skew tensors and $[\Delta]$ is an arbitrary diagonal matrix, we obtain the system of Euler–Lagrange equations

$$\frac{d}{dt}\left(R_L \frac{\partial L^D}{\partial \Omega_L} R_L^T \right) = \text{sk}\left(\frac{\partial L^D}{\partial R_L} R_L^T \right),$$

$$\frac{d}{dt}\left(\frac{\partial L^D}{\partial \dot{D}} \right) = \frac{\partial L^D}{\partial D}, \tag{5.2.51}$$

$$\frac{d}{dt}\left(R_R^T \frac{\partial L^D}{\partial \Omega_R} R_R \right) = \text{sk}\left(R_R^T \frac{\partial L^D}{\partial R_R} \right).$$

It is worth emphasizing that each of these tensor-valued equations is equivalent to three scalar equations.

For a pseudo-rigid body, (5.2.43), (5.1.12), (5.1.17), and (5.1.33) give the explicit formula

$$L^D = \tfrac{1}{2}(\dot{D} + \Omega_L D + D\Omega_R)R_R E_R R_R^T \cdot (\dot{D} + \Omega_L D + D\Omega_R)$$

$$- \sigma(DR_R) + u(R_L DR_R), \tag{5.2.52}$$

where we have eliminated R_L from σ by the principle of material frame indifference. In special cases, this expression simplifies further. If the Euler tensor E_R is a multiple of I, indicating that the mass distribution is completely symmetric, then R_R drops out of the first term. If the material is isotropic, so that $\sigma(F) = \sigma(FQ)$ for any $Q \in \mathcal{O}(\mathcal{V})$, then R_R drops out of the second term. Finally, suppose the load potential has the property that

$$u(R_L FR_R) = u(F) \tag{5.2.53}$$

for any $R_L, R_R \in \mathcal{O}(\mathcal{V})$. In this case, both R_L and R_R are absent from the last term in L^D. We conclude that L^D is independent of both rotations, and so $(5.2.51)_{1,3}$ give us

$$R_L \frac{\partial L^D}{\partial \Omega_L} R_L^T = C_L, \qquad R_R^T \frac{\partial L^D}{\partial \Omega_R} R_R = C_R, \qquad (5.2.54)$$

where C_L and C_R are constant skew tensors. These give us six conservation laws. The three conventional laws of conservation of angular momentum are functions of these six. In addition, we obtain three others that reflect the invariance owing to material isotropy and the symmetric distribution of mass.

The identity (5.2.53) implies that the net torque on the body vanishes, but we can conclude much more. By standard reduction theorems for constitutive relations, we conclude from (5.2.53) that u is a function only of the principal invariants of F. Consequently M, as given by $(5.1.59)_1$, has a representation the same as that given in $(2.7.15)_1$. In particular, since F in (5.2.53) is the tensor D, we conclude that M is diagonal with respect to the principal axes of D. Thus, *the conservation laws* (5.2.54) *apply provided the external force-moment represents a set of stretches or compressions along the principal axes of strain.*

A restricted version of $(5.2.54)_2$ is valid if the Euler tensor has an axis of symmetry and if σ and u are transversely isotropic along this same axis. The analysis of this case is left to the reader (cf. COHEN and MUNCASTER [1984]).

(iii) The Stability of Steady Motions: A Special Case

In both mass-point and rigid-body mechanics, the variational framework is particularly useful for the analysis of small oscillations. While generally this means oscillations about an equilibrium configuration, the case of small motions about a steady state has also been examined, and it is this latter case that arises naturally in a class of problems for pseudo-rigid bodies. Roto-deformations, which we have treated in detail in Section 4(ii), are motions in which rotation and deformation exhibit a strong interaction. They make up one of the applications most ideally suited to the theory of pseudo-rigid bodies, and in the context of steady-state motions we have been able to analyze them in considerable detail. Here we pose the additional question: *Are steady roto-deformations stable?*

Steady roto-deformations, defined by (4.2.1), (4.2.2), and (4.2.8), may be characterized as motions in which the stretch tensor U and the angular velocity w are constant. Thus, a steady roto-deformation, when analyzed in terms of U and Euler angles θ, φ, and ψ, represents a *steady state*, but the same deformation analyzed in terms of U and the components of angular velocity w_1, w_2, and w_3 represents an *equilibrium state*. This observation provides two approaches to the question of stability, each with its own advantages. In this section, we consider an analysis of small motions about equilibrium based on U and w; Section 5(iv) addresses small motions about a steady state in terms of U and the Euler angles. Both analyses are based on linearized stability criteria.

Within rigid-body mechanics, the conventional analysis of the stability of steady rotation based on EULER's equations is formulated in terms of the components of angular velocity. Our objective here is a comparable analysis for pseudo-rigid bodies. In order to strengthen the analogy and also keep the analysis simple, we examine only a special case of the stability question. Namely, we consider roto-deformations of the form

$$F = RD, \qquad (5.3.1)$$

in which D is positive-definite and has a diagonal representation relative to the principal axes of inertia:

$$[D] = \text{diag}(u_1, u_2, u_3),$$
$$[E_R] = \text{diag}(E_1, E_2, E_3). \qquad (5.3.2)$$

Let the fixed basis vectors D_i be chosen to coincide with the principal axes of E_R, so that the pure deformation D consists of stretches along these axes. In this section, we examine only steady motions of this type and also stability with respect to small perturbations of the same form.

Assume that the load potential u is absent. Then (5.2.25) shows that the Lagrangian has the form

$$L^U = \tfrac{1}{2}(\dot{D} + \Omega D)E_R \cdot (\dot{D} + \Omega D) - \sigma(D). \qquad (5.3.3)$$

As (5.2.36) indicates, Ω is determined by the components $\hat{w}_1$, $\hat{w}_2$, and $\hat{w}_3$ of its axial vector ω relative to the vectors D_i, these being also the components of w relative to corotating body axes $\hat{d}_i = RD_i$. For notational convenience, in the remainder of this section we drop the superscript U from L^U and the hat from $\hat{w}_i$, and the summation convention applies. Then with (5.3.2) and (5.3.3) we can write the Lagrangian in the component form

$$L(u_i, \dot{u}_i, w_i) = \tfrac{1}{2}(E_i \dot{u}_i^2 + E_i u_i^2(w^2 - w_i^2)) - \sigma^s(u_i), \qquad (5.3.4)$$

where $w^2 = w_i w_i$, and (2.7.22) has been used to express the stored energy as a function σ^s of the principal stretches. The Euler–Lagrange equations associated with L follow from $(5.2.23)_2$ and (5.2.33):

$$\frac{d}{dt}\left(\frac{\partial L}{\partial \dot{u}_i}\right) = \frac{\partial L}{\partial u_i},$$
$$\qquad\qquad\qquad\qquad i = 1, 2, 3. \qquad (5.3.5)$$
$$\frac{d}{dt}\left(\frac{\partial L}{\partial w_i}\right) = -\varepsilon_{ijk} w_j \frac{\partial L}{\partial w_k},$$

A steady roto-deformation is an equilibrium point of this system. Denoting an equilibrium by the superscript 0, we obtain from (5.3.5) the governing equations

$$\frac{\partial L}{\partial u_i}(u_i^0, 0, w_i^0) = 0, \qquad \varepsilon_{ijk} w_j^0 \frac{\partial L}{\partial w_k}(u_i^0, 0, w_i^0) = 0. \qquad (5.3.6)$$

The second of these shows that

$$\frac{\partial L}{\partial w_k}(u_i^0, 0, w_i^0) = \lambda w_k^0, \qquad k = 1, 2, 3, \tag{5.3.7}$$

for some scalar λ. By (5.3.4) this system has the explicit form

$$w_1^0(\lambda - E_2 u_2^{02} - E_3 u_3^{02}) = 0,$$
$$w_2^0(\lambda - E_1 u_1^{02} - E_3 u_3^{02}) = 0, \tag{5.3.8}$$
$$w_3^0(\lambda - E_1 u_1^{02} - E_2 u_2^{02}) = 0.$$

Except for degenerate cases, only one of the terms in parentheses here can vanish. The corresponding component of angular velocity is arbitrary, and all other components vanish. By reordering indices, we may assume that

$$w_1^0 = \omega, \qquad w_2^0 = 0, \qquad w_3^0 = 0, \qquad \lambda = E_2 u_2^{02} + E_3 u_3^{02}, \tag{5.3.9}$$

where ω is arbitrary. Then $(5.3.6)_1$ reduces to the system

$$\frac{\partial \sigma^s}{\partial u_1}(u_i^0) = 0,$$

$$\frac{\partial \sigma^s}{\partial u_2}(u_i^0) = \omega^2 E_2 u_2^0, \tag{5.3.10}$$

$$\frac{\partial \sigma^s}{\partial u_3}(u_i^0) = \omega^2 E_3 u_3^0.$$

These are precisely the conditions $(4.2.13)_{1-3}$ from our initial study of steady roto-deformations.

In order to draw comparisons with results from rigid-body mechanics, we write the Lagrangian $(5.2.38)_2$ for a rigid body in terms of components. Let A, B, and C denote the principal moments of inertia relative to body axes. Equivalently, these are the moments of inertia relative to the basis D_i in the reference placement, and therefore $[J_R] = \mathrm{diag}(A, B, C)$. Since ω has components w_1, w_2, w_3 relative to this basis, (5.2.38) becomes

$$L = \tfrac{1}{2}A w_1^2 + \tfrac{1}{2}B w_2^2 + \tfrac{1}{2}C w_3^2. \tag{5.3.11}$$

The equilibrium equations in this case are still given by $(5.3.6)_2$, or by their equivalents (5.3.7), and these agree precisely with (5.3.8) provided we set

$$A = E_2 u_2^{02} + E_3 u_3^{02},$$
$$B = E_1 u_1^{02} + E_3 u_3^{02}, \tag{5.3.12}$$
$$C = E_1 u_1^{02} + E_2 u_2^{02}.$$

The generic solution (5.3.9) then shows that *a steady spinning pseudo-rigid motion is possible about any principal axis of inertia*, a result paralleling that for rigid motions that we have already obtained in Section 4(i).

In order to study small motions superimposed on the preceding equilibrium state, we develop an appropriate linearization of the system (5.3.5). First, we note that, with the identifications (5.3.12), (5.3.7) takes the explicit form

$$\frac{\partial L}{\partial w_1}\bigg|^0 = A\omega, \qquad \frac{\partial L}{\partial w_2}\bigg|^0 = 0, \qquad \frac{\partial L}{\partial w_3}\bigg|^0 = 0, \tag{5.3.13}$$

in both the rigid and pseudo-rigid cases. Considering first the rigid-body case, let us introduce the perturbations

$$w_1 = \omega + \zeta_1, \qquad w_2 = \zeta_2, \qquad w_3 = \zeta_3. \tag{5.3.14}$$

Then

$$\varepsilon_{ijk} w_j \frac{\partial L}{\partial w_k} = \varepsilon_{ijk} w_j^0 \frac{\partial L}{\partial w_k} + \varepsilon_{ijk} \zeta_j \frac{\partial L}{\partial w_k}\bigg|^0 + \cdots, \tag{5.3.15}$$

where the dots denote terms that are nonlinear in ζ_i. Let L' denote the approximation to the rigid-body Lagrangian L, which is quadratic in the variables ζ_i. Then (5.3.13) and (5.3.15) show that the linearized equations corresponding to (5.3.5) are

$$\frac{d}{dt}\left(\frac{\partial L'}{\partial \zeta_1}\right) = 0, \qquad \frac{d}{dt}\left(\frac{\partial L'}{\partial \zeta_2}\right) = \omega\left(\frac{\partial L'}{\partial \zeta_3} - A\zeta_3\right),$$

$$\frac{d}{dt}\left(\frac{\partial L'}{\partial \zeta_3}\right) = -\omega\left(\frac{\partial L'}{\partial \zeta_2} - A\zeta_2\right). \tag{5.3.16}$$

By placing (5.3.14) into (5.3.11) and expanding the result to within quadratic terms in ζ_i, we find that L' is the function

$$L' = \tfrac{1}{2}A\omega^2 + A\omega\zeta_1 + \tfrac{1}{2}A\zeta_1^2 + \tfrac{1}{2}B\zeta_2^2 + \tfrac{1}{2}C\zeta_3^2. \tag{5.3.17}$$

Then $(5.3.16)_{2,3}$ become

$$B\dot{\zeta}_2 = \omega(C - A)\zeta_3, \qquad C\dot{\zeta}_3 = -\omega(B - A)\zeta_2. \tag{5.3.18}$$

Eliminating ζ_3, we obtain the second-order equation

$$\ddot{\zeta}_2 + \omega^2 \frac{(C - A)(B - A)}{BC}\zeta_2 = 0, \tag{5.3.19}$$

and standard results on differential equations now show that the steady spinning motion of a rigid body is stable if and only if

$$(C - A)(B - A) > 0. \tag{5.3.20}$$

This gives the classical result[4] that *a steady spinning motion of a rigid*

---

[4] Cf. SYMON [1960, p. 455].

*body is stable if the axis of rotation corresponds either to the largest or to
the smallest moment of inertia.*

The analysis of stability for a pseudo-rigid body proceeds in the same
way. Let us introduce, in addition to (5.3.14), the perturbations

$$\delta_1 = (u_1 - u_1^0)/u_1^0, \qquad \delta_2 = (u_2 - u_2^0)/u_2^0, \qquad \delta_3 = (u_3 - u_3^0)/u_3^0. \quad (5.3.21)$$

Placing these and (5.3.14) into L and expanding to within quadratic
terms in δ_i and ζ_i, we obtain an approximate Lagrangian L'. Then the
linear equations corresponding to (5.3.5) are (5.3.16) and the system

$$\frac{d}{dt}\left(\frac{\partial L'}{\partial \dot{\delta}_i}\right) = \frac{\partial L'}{\partial \delta_i}, \qquad i = 1, 2, 3. \tag{5.3.22}$$

In order to find L', it is useful to introduce the Euler tensor E^0 of the
steady state, namely,

$$E^0 = D^0 E_R D^0,$$
$$[E^0] = \mathrm{diag}(E_1^0, E_2^0, E_3^0) = \mathrm{diag}(E_1 u_1^{02}, E_2 u_2^{02}, E_3 u_3^{02}). \tag{5.3.23}$$

By comparison with (5.3.12) we see that $A = E_2^0 + E_3^0$, $B = E_1^0 + E_3^0$,
and $C = E_1^0 + E_2^0$, and so the inertia tensor J_R for the associated rigid
motion corresponds via (2.1.21) to the Euler tensor E^0. The approximate
Lagrangian L' now follows from (5.3.4) by no more than elementary
calculation. It has the form

$$\begin{aligned}
L' = {}&\tfrac{1}{2}A\omega^2 + A\omega\zeta_1 + \tfrac{1}{2}A\zeta_1^2 + \tfrac{1}{2}B\zeta_2^2 + \tfrac{1}{2}C\zeta_3^2 \\
&+ \tfrac{1}{2}E_i^0\dot{\delta}_i^2 + \tfrac{1}{2}\omega^2 E_2^0\delta_2^2 + \tfrac{1}{2}\omega^2 E_3^0\delta_3^2 \\
&+ 2\omega\zeta_1(E_2^0\delta_2 + E_3^0\delta_3) - \sigma^0 - \tfrac{1}{2}\sigma_{ij}^0\delta_i\delta_j,
\end{aligned} \tag{5.3.24}$$

where we have used (5.3.10) to eliminate the first derivatives of σ^s, and
we have set

$$\sigma^0 = \sigma^s(u_i^0), \qquad \sigma_{jk}^0 = u_j^0 u_k^0 \frac{\partial^2 \sigma^s}{\partial u_j \partial u_k}(u_i^0) \tag{5.3.25}$$

(no sum on j and k). The approximate Lagrangian (5.3.17) for a rigid
body appears explicitly here as the first part of L'. The remaining terms
arise from the interaction between rotation and deformation. It is important to note that these extra terms do not contain ζ_2 and ζ_3. As a
consequence, (5.3.18) and (5.3.19) apply also to a pseudo-rigid body: *rigid
bodies and pseudo-rigid bodies exhibit the same linearized equations of
motion for the components of angular velocity perpendicular to the axis of
steady motion.* This is a surprising result. As we shall see in Section 5(iv),
it is principally a consequence of the class of deformations we have
chosen to consider, namely, those representing extensions or compressions
along the principal axes of inertia.

It should not be concluded that the conditions for instability for a pseudo-rigid body are the same as those for a rigid body. Certainly, we require (5.3.20) once again, but the interpretation of this inequality is different. Recall that A, B, and C are the principal moments of inertia in the steady state. This means that each is a function of the angular speed ω, and therefore (5.3.20) must be written more explicitly as

$$(C(\omega) - A(\omega))(B(\omega) - A(\omega)) > 0. \tag{5.3.26}$$

It is possible that this condition might be valid for small values of ω and yet fail for larger values, especially since there is generally a contraction of the body, increasing with spin rate, along the axis of spin.

The case of a St. Vénant–Kirchhoff material provides an explicit example. The squares of the equilibrium values u_1^0, u_2^0, and u_3^0 are then given by the right-hand sides of (4.2.15). Since $C - A = E_1 u_1^{02} - E_3 u_3^{02}$, we see that $C(\hat{\omega}) = A(\hat{\omega})$ at the critical angular speed $\hat{\omega}$ satisfying

$$E_1 - \hat{\omega}^2 \beta E_1 = E_3 + \frac{1}{\mu}\hat{\omega}^2 E_3^2 - \hat{\omega}^2 \beta E_3, \tag{5.3.27}$$

where

$$\beta = \frac{\lambda(E_2 + E_3)}{\mu(3\lambda + 2\mu)}. \tag{5.3.28}$$

Similarly, $B - A = E_1 u_1^{02} - E_2 u_2^{02}$, and so $B(\tilde{\omega}) = A(\tilde{\omega})$ provided $\tilde{\omega}$ satisfies

$$E_1 - \tilde{\omega}^2 \beta E_1 = E_2 + \frac{1}{\mu}\tilde{\omega}^2 E_2^2 - \tilde{\omega}^2 \beta E_2. \tag{5.3.29}$$

These give us

$$\hat{\omega}^2 = \frac{\mu}{\mu\beta + E_3^2/(E_1 - E_3)}, \qquad \tilde{\omega}^2 = \frac{\mu}{\mu\beta + E_2^2/(E_1 - E_2)}. \tag{5.3.30}$$

If $E_1 > E_2 > E_3$, a simple calculation shows that $0 < \tilde{\omega} < \hat{\omega}$. Moreover, $C(0) - A(0) = E_1 - E_3 > 0$ and $B(0) - A(0) = E_1 - E_2 > 0$. Thus, (5.3.26) is valid for $0 \le \omega < \tilde{\omega}$ and for $\hat{\omega} < \omega$, and it is violated for $\tilde{\omega} < \omega < \hat{\omega}$. Therefore, the moment of inertia $A(\omega)$ increases as the rate of spinning increases, beginning as the smallest moment of inertia in the steady state, then changing to that of intermediate size, and finally becoming the largest. Clearly the ability of the pseudo-rigid body to deform is the principal factor giving rise to the intermediate range of instability.

Finally, consider the linear equations (5.3.22). Using (5.3.24) we can write these explicitly in the form

$$E_1^0 \ddot{\delta}_1 + \sigma_{1j}^0 \delta_j = 0,$$
$$E_2^0 \ddot{\delta}_2 + \sigma_{2j}^0 \delta_j = \omega^2 E_2^0 \delta_2 + 2\omega E_2^0 \zeta_1, \tag{5.3.31}$$
$$E_3^0 \ddot{\delta}_3 + \sigma_{3j}^0 \delta_j = \omega^2 E_3^0 \delta_3 + 2\omega E_3^0 \zeta_1.$$

By $(5.3.16)_1$ and $(5.3.24)$,

$$A\omega + A\zeta_1 + 2\omega E_2^0 \delta_2 + 2\omega E_3^0 \delta_3 = p \qquad (5.3.32)$$

for some constant p. This constant is a conserved generalized momentum for the motion, and it is useful to give it the value in the steady state, namely, $p = A\omega$. Then we obtain

$$\zeta_1 = -2\omega \frac{E_2^0}{A} \delta_2 - 2\omega \frac{E_3^0}{A} \delta_3. \qquad (5.3.33)$$

Substituting this into $(5.3.31)$ and simplifying, we obtain a system that can be written in the tensor form

$$E^0 \delta + \Sigma^0 \delta = \frac{\omega^2}{A} E^0 K E^0 \delta, \qquad (5.3.34)$$

where $[\delta] = (\delta_i)$, $[\Sigma^0] = (\sigma_{ij}^0)$, and

$$[K] = \begin{bmatrix} 0 & 0 & 0 \\ 0 & (E_3^0/E_2^0) - 3 & -4 \\ 0 & -4 & (E_2^0/E_3^0) - 3 \end{bmatrix}. \qquad (5.3.35)$$

The stability characteristics of the rotational equations $(5.3.18)$ depend crucially on the relative magnitudes of the steady-state moments of inertia, and consequently on the fact that these moments of inertia change with the value of ω. Similar but more complex issues arise in connection with $(5.3.34)$. The tensor Σ^0 carries information about material response and it is important to make certain physically based assumptions about this response in order to proceed. We note first that Σ^0, E^0, K, and A are all functions of the angular speed ω of the steady state. The components of Σ^0 represent the "elasticities" associated with linearization about the steady state. If we require that the stored energy σ attain a local minimum at the steady state, then for each ω the tensor $\Sigma^0(\omega)$ should be positive-definite and symmetric.[5] The actual stability conditions for $(5.3.34)$ come from consideration of the tensor $C(\omega)$ given by

$$C(\omega) = \Sigma^0(\omega) - \frac{\omega^2}{A(\omega)} E^0(\omega) K(\omega) E^0(\omega). \qquad (5.3.36)$$

If $K(\omega)$ were negative-definite, then $C(\omega)$ would be positive-definite, and stability would follow by a result of SYNGE [1960]. Unfortunately, this is not the case. A simple calculation shows that K has both a positive and a negative eigenvalue, and hence it is indefinite. Certainly, $C(0) = \Sigma^0(0)$, which by assumption is positive-definite, and so for ω sufficiently small

---

[5] This is the analogue here of the strong-ellipticity condition for elastic bodies (cf. TRUESDELL and NOLL [1965, Eq. (68b.2)]).

the solutions of (5.3.34) are stable. However, for ω large we can draw no general conclusion. While the second term in $C(\omega)$ becomes important, it may happen that $\Sigma^0(\omega)$ changes in such a manner that $C(\omega)$ remains positive-definite. Conclusions of this type would require a deeper analysis of material response than we have available here.

The St. Vénant–Kirchhoff material provides some insight into the subtleties of this problem. From (2.7.23) and (2.7.33)$_1$ we see that

$$\frac{\partial \sigma^s}{\partial u_i} = T_i = \tfrac{1}{2}(\lambda(u_1^2 + u_2^2 + u_3^2) + 2\mu u_i^2 - (3\lambda + 2\mu))u_i, \qquad (5.3.37)$$

where the summation convention does not apply in this case. Therefore,

$$\frac{\partial^2 \sigma^s}{\partial u_i \partial u_j} = \tfrac{1}{2}(\lambda(u_1^2 + u_2^2 + u_3^2) + 2\mu u_i^2 - (3\lambda + 2\mu))\delta_{ij}$$

$$+ \lambda u_i u_j + 2\mu u_i u_j \delta_{ij}. \qquad (5.3.38)$$

In equilibrium the term in parentheses can be calculated directly from (4.2.15). As a result, the coefficients σ_{ij}^0 given by (5.3.25) become

$$\sigma_{11}^0 = (\lambda + 2\mu)u_1^{04}, \qquad \sigma_{ij}^0 = \lambda u_i^{02} u_j^{02} \quad \text{for } i \neq j,$$
$$\sigma_{22}^0 = (\lambda + 2\mu)u_2^{04} + \omega^2 E_2^0, \qquad \sigma_{33}^0 = (\lambda + 2\mu)u_3^{04} + \omega^2 E_3^0. \qquad (5.3.39)$$

These can be written in the tensor form

$$\Sigma^0 = 2\mu D^{04} + \lambda D^{02} P D^{02} + \frac{\omega^2}{A} E^0 M E^0, \qquad (5.3.40)$$

where $[P]$ is the matrix whose entries all equal 1,

$$[M] = \text{diag}\left(0, 1 + \frac{E_3^0}{E_2^0}, 1 + \frac{E_2^0}{E_3^0}\right), \qquad (5.3.41)$$

and we have used the fact that $A = E_2^0 + E_3^0$. Placing this in (5.3.36), we obtain

$$C(\omega) = 2\mu D^{04} + \lambda D^{02} P D^{02} + \frac{4\omega^2}{A} E^0 N E^0, \qquad (5.3.42)$$

where

$$[N] = \begin{bmatrix} 0 & 0 & 0 \\ 0 & 1 & 1 \\ 0 & 1 & 1 \end{bmatrix}. \qquad (5.3.43)$$

The term $2\mu D^{04}$ is positive-definite. Moreover, an examination of the eigenvalues of P and N shows that each of these matrices is positive-semidefinite. We have seen in connection with the steady solution (4.2.15) that for the St. Vénant–Kirchhoff material it is reasonable to assume that $\lambda > 0$. Then (5.3.42) shows that $C(\omega)$ is positive-definite for all ω for which the steady solution is valid. Thus, (5.3.34) exhibits only stable motions in this case. Only by examining the elasticities σ_{ij}^0 explicitly have we been able to show that there arises a term, the last in (5.3.40), that compensates for the indefinite character of the second term in (5.3.36) contributing to $C(\omega)$. Whether a similar term arises for more general materials is a matter for further research.

(iv) Ignorable Coordinates: Remarks on the General Analysis of Stability

The preceding analysis of stability, phrased in terms of the components of angular velocity, provides for pseudo-rigid bodies a natural parallel of calculations now commonplace in the mechanics of rigid bodies. An alternate approach, more useful for some applications, comes from representing the spinning of a body in terms of a set of Euler angles. Within a variational setting, this leads to an analysis of stability directly in terms of a perturbation of the Lagrangian about a steady motion. The general theory behind this type of analysis is well known (cf. PARS [1965]). It centers on the idea that one of the generalized coordinates is *ignorable*; that is, only its time derivative enters the Lagrangian, and as a consequence the associated generalized momentum is a constant of the motion. By eliminating this information from the Lagrangian L, we obtain a new function R called the *Routhian*, and then the stability of the steady motion for L can be analyzed in terms of the stability of an equilibrium state for R.

For pseudo-rigid bodies, an analysis of the general stability of steady spinning following the above approach is a formidable task. Here we consider a more modest goal. In order to illustrate the variational analysis of motion near a steady state, we examine in detail the stability of steady spinning of a rigid body. Surprisingly, we have not found this analysis in the literature. Second, we present the construction and approximation of the Routhian of a pseudo-rigid body for general motions near a steady spinning motion. This calculation serves an important role. We have seen in Section 5(iii) that for a special class of perturbations the linearized equations decouple into two smaller systems, one being exactly that appearing in rigid-body mechanics. As a consequence, we were able to carry over the conditions for instability of a spinning rigid body, suitably reinterpreted, to pseudo-rigid bodies. The approximation we give for the Routhian for general perturbations will allow us to see why this split occurs, and whether the general case exhibits a similar decoupling.

Consider a rigid body subject to no external loads. The Lagrangian then has the form (5.3.11) relative to body axes. It is convenient to work in terms of Euler angles φ_1, φ_2, and φ_3 measuring rotations about three different axes. These angles are defined by (4.1.51), and in terms of them the components of angular velocity relative to body axes are given by (4.1.52). Substituting these component expressions into (5.3.11) and simplifying, we obtain

$$L = \tfrac{1}{2}\dot{\varphi}_1^2(A \cos^2 \varphi_2 \cos^2 \varphi_3 + B \cos^2 \varphi_2 \sin^2 \varphi_3 + C \sin^2 \varphi_2)$$
$$+ \tfrac{1}{2}\dot{\varphi}_2^2(A \sin^2 \varphi_3 + B \cos^2 \varphi_3) + \tfrac{1}{2}C\dot{\varphi}_3^2$$
$$+ \dot{\varphi}_1((A - B)\dot{\varphi}_2 \cos \varphi_2 \cos \varphi_3 \sin \varphi_3 + C\dot{\varphi}_3 \sin \varphi_2). \tag{5.4.1}$$

The angle φ_1 is absent and so represents an ignorable coordinate. Consequently, the Euler–Lagrange equation corresponding to φ_1 can be integrated to give

$$p = \frac{\partial L}{\partial \dot{\varphi}_1} = \dot{\varphi}_1(A \cos^2 \varphi_2 \cos^2 \varphi_3 + B \cos^2 \varphi_2 \sin^2 \varphi_3 + C \sin^2 \varphi_2)$$

$$+ (A - B)\dot{\varphi}_2 \cos \varphi_2 \cos \varphi_3 \sin \varphi_3 + C\dot{\varphi}_3 \sin \varphi_2, \tag{5.4.2}$$

where p is a constant.

Since φ_1 is ignorable, a steady spinning motion is characterized by $\dot{\varphi}_1$, φ_2, and φ_3 being constants. Naturally, the Euler–Lagrange equations must be satisfied. One of these gives us (5.4.2). The other two become

$$0 = \frac{\partial L}{\partial \varphi_2}$$
$$= \dot{\varphi}_1^2 \sin \varphi_2 \cos \varphi_2(C - A \cos^2 \varphi_3 - B \sin^2 \varphi_3)$$
$$+ \dot{\varphi}_1(C\dot{\varphi}_3 \cos \varphi_2 + (B - A)\dot{\varphi}_2 \sin \varphi_2 \cos \varphi_3 \sin \varphi_3),$$
$$\tag{5.4.3}$$
$$0 = \frac{\partial L}{\partial \varphi_3}$$
$$= \dot{\varphi}_1^2(B - A) \cos^2 \varphi_2 \sin \varphi_3 \cos \varphi_3 + \dot{\varphi}_2^2(A - B) \sin \varphi_3 \cos \varphi_3$$
$$+ \dot{\varphi}_1 \dot{\varphi}_2(A - B) \cos \varphi_2(\cos^2 \varphi_3 - \sin^2 \varphi_3).$$

Since $\dot{\varphi}_2$ and $\dot{\varphi}_3$ both vanish, one solution of these equations is

$$\dot{\varphi}_1^0 = p/A, \qquad \varphi_2^0 = 0, \qquad \varphi_3^0 = 0. \tag{5.4.4}$$

These define a steady-state rotation about the axis D_1 corresponding to the principal moment of inertia A.

In order to analyze the stability of this steady motion, we first eliminate φ_1 from the problem. This is done by solving (5.4.2) for $\dot{\varphi}_1$ and substituting the result in the Lagrangian L. It is useful to carry this calculation one step further by forming the *Routhian*

$$R = L - \dot{\varphi}_1 p; \tag{5.4.5}$$

then the differential equations for φ_2 and φ_3, with φ_1 eliminated, are precisely the Euler–Lagrange equations for R (cf. PARS [1965]). The linearized equations for the analysis of stability then follow from a quadratic approximation of R about $\varphi_2^0 = \varphi_3^0 = 0$.

Set

$$J = A \cos^2 \varphi_2 \cos^2 \varphi_3 + B \cos^2 \varphi_2 \sin^2 \varphi_3 + C \sin^2 \varphi_2. \quad (5.4.6)$$

Then (5.4.2) gives us

$$\dot{\varphi}_1 = J^{-1}[p - (A - B)\dot{\varphi}_2 \cos \varphi_2 \cos \varphi_3 \sin \varphi_3 - C\dot{\varphi}_3 \sin \varphi_2]. \quad (5.4.7)$$

The Lagrangian can be written in the preliminary form

$$L = -\tfrac{1}{2}J\dot{\varphi}_1^2 + p\dot{\varphi}_1 + \tfrac{1}{2}\dot{\varphi}_2^2(A \sin^2 \varphi_3 + B \cos^2 \varphi_3) + \tfrac{1}{2}C\dot{\varphi}_3^2. \quad (5.4.8)$$

Placing this in (5.4.5) and then eliminating $\dot{\varphi}_1$ by use of (5.4.7), we obtain the Routhian

$$R = \tfrac{1}{2}\dot{\varphi}_2^2(A \sin^2 \varphi_3 + B \cos^2 \varphi_3) + \tfrac{1}{2}C\dot{\varphi}_3^2$$

$$-\tfrac{1}{2}J^{-1}[p - (A - B)\dot{\varphi}_2 \cos \varphi_2 \cos \varphi_3 \sin \varphi_3 - C\dot{\varphi}_3 \sin \varphi_2]^2. \quad (5.4.9)$$

In order to develop an approximation of R that is quadratic in φ_2, φ_3, $\dot{\varphi}_2$, and $\dot{\varphi}_3$, we note that

$$R = \tfrac{1}{2}B\dot{\varphi}_2^2 + \tfrac{1}{2}C\dot{\varphi}_3^2 - \frac{p^2}{2J} + \frac{p}{J}(A - B)\dot{\varphi}_2\varphi_3 + \frac{p}{J}C\dot{\varphi}_3\varphi_2 + \cdots, \quad (5.4.10)$$

where the dots denote higher-order terms. To complete the calculation, we need a corresponding approximation for J. By (5.4.6) we see that

$$J = A(1 - \sin^2 \varphi_2)(1 - \sin^2 \varphi_3) + B(1 - \sin^2 \varphi_2) \sin^2 \varphi_3 + C \sin^2 \varphi_2$$

$$= A - A\varphi_3^2 - A\varphi_2^2 + B\varphi_3^2 + C\varphi_2^2 + \cdots$$

$$= A + (C - A)\varphi_2^2 + (B - A)\varphi_3^2 + \cdots. \quad (5.4.11)$$

Thus,

$$\frac{1}{J} = \frac{1}{A}\left(1 + \frac{A - C}{A}\varphi_2^2 + \frac{A - B}{A}\varphi_3^2\right) + \cdots. \quad (5.4.12)$$

Placing this into (5.4.10), we obtain the quadratic approximation R' to the Routhian in the form

$$R' = \tfrac{1}{2}B\dot{\varphi}_2^2 + \tfrac{1}{2}C\dot{\varphi}_3^2 + \frac{p(A - B)}{A}\dot{\varphi}_2\varphi_3 + \frac{pC}{A}\dot{\varphi}_3\varphi_2$$

$$-\frac{p^2}{2A^2}(A + (A - C)\varphi_2^2 + (A - B)\varphi_3^2). \quad (5.4.13)$$

The linearized equations for motions near the steady state are

$$\frac{d}{dt}\left(\frac{\partial R'}{\partial \dot{\varphi}_i}\right) = \frac{\partial R'}{\partial \varphi_i}, \qquad i = 2, 3. \quad (5.4.14)$$

In view of (5.4.13), these take the explicit form

$$B\ddot{\varphi}_2 + \frac{p}{A}(A - B - C)\dot{\varphi}_3 + \frac{p^2}{A^2}(A - C)\varphi_2 = 0,$$

$$C\ddot{\varphi}_3 - \frac{p}{A}(A - B - C)\dot{\varphi}_2 + \frac{p^2}{A^2}(A - B)\varphi_3 = 0. \tag{5.4.15}$$

The solutions of this system are proportional to $\exp i\xi t$, or are linear combinations of such terms, if ξ is any solution of the characteristic equation

$$\det\begin{bmatrix} -B\xi^2 + \dfrac{p^2}{A^2}(A - C) & i\xi\dfrac{p}{A}(A - B - C) \\[3mm] -i\xi\dfrac{p}{A}(A - B - C) & -C\xi^2 + \dfrac{p^2}{A^2}(A - B) \end{bmatrix} = 0. \tag{5.4.16}$$

An elementary calculation gives us

$$\left(\frac{A\xi}{p}\right)^2 = \begin{cases} \dfrac{(A - C)(A - B)}{BC}, \\[3mm] 1. \end{cases} \tag{5.4.17}$$

We conclude, as in Section 5(iii), that the steady spinning motion of a rigid body is stable if and only if A, B, and C satisfy the restriction (5.3.20).

In order to adapt the preceding analysis to the case of a pseudo-rigid body, we represent the general deformation F in the form (5.2.41) with R_L determined by Euler angles φ_1, φ_2, φ_3 and R_R determined by Euler angles θ_1, θ_2, and θ_3. More specifically, in terms of the general orthogonal tensor Q_i defined by (4.1.11), we set

$$F = R_L D R_R,$$

$$R_L = Q_1(\varphi_1)Q_2(\varphi_2)Q_3(\varphi_3), \tag{5.4.18}$$

$$R_R = Q_3(\theta_3)Q_2(\theta_2)Q_1(\theta_1).$$

Since $(d/dt)Q_i(\alpha) = \dot{\alpha}Q_i(\alpha)A_i = \dot{\alpha}A_iQ_i(\alpha)$ for each i (cf. Section 4(i)), we find that

$$\dot{R}_L = R_L(\dot{\varphi}_i\hat{A}_i), \qquad \dot{R}_R = (\dot{\theta}_i\tilde{A}_i)R_R, \tag{5.4.19}$$

where

$$\hat{A}_1 = Q_3^T(\varphi_3)Q_2^T(\varphi_2)A_1 Q_2(\varphi_2)Q_3(\varphi_3),$$

$$\hat{A}_2 = Q_3^T(\varphi_3)A_2 Q_3(\varphi_3), \qquad \hat{A}_3 = A_3,$$

$$\tilde{A}_1 = Q_3(\theta_3)Q_2(\theta_2)A_1 Q_2^T(\theta_2)Q_3^T(\theta_3), \tag{5.4.20}$$

$$\tilde{A}_2 = Q_3(\theta_3)A_2 Q_3^T(\theta_3), \qquad \tilde{A}_3 = A_3.$$

In (5.4.19) and henceforth the summation convention applies. It follows then from (5.4.19) that

$$\dot{F} = R_{\text{L}}(\dot{D} + R_{\text{L}}^{\text{T}}\dot{R}_{\text{L}}D + D\dot{R}_{\text{R}}R_{\text{R}}^{\text{T}})R_{\text{R}}$$

$$= R_{\text{L}}(\dot{D} + \dot{\varphi}_i\hat{A}_iD + \dot{\theta}_iD\tilde{A}_i)R_{\text{R}}. \qquad (5.4.21)$$

If we consider only an isotropic body, so that $\sigma(F) = \sigma^{\text{s}}(D)$, and assume that the load potential u is absent, then the Lagrangian $(5.2.25)_1$ becomes

$$L = \tfrac{1}{2}(\dot{D} + \dot{\varphi}_i\hat{A}_iD + \dot{\theta}_iD\tilde{A}_i)E \cdot (\dot{D} + \dot{\varphi}_j\hat{A}_jD + \dot{\theta}_jD\tilde{A}_j) - \sigma^{\text{s}}(D), \quad (5.4.22)$$

where

$$E = Q_3(\theta_3)Q_2(\theta_2)Q_1(\theta_1)E_{\text{R}}Q_1^{\text{T}}(\theta_1)Q_2^{\text{T}}(\theta_2)Q_3^{\text{T}}(\theta_3). \qquad (5.4.23)$$

Each of the skew tensors $\hat{A}_i$ and $\tilde{A}_i$ and the symmetric tensor E are independent of φ_1. Therefore, φ_1 is once again an ignorable coordinate. It is useful to set

$$K = \dot{D} + \dot{\varphi}_\alpha\hat{A}_\alpha D + \dot{\theta}_iD\tilde{A}_i \qquad (5.4.24)$$

in which α takes values of 2 and 3, and the summation convention applies. Then the Lagrangian has the simpler form

$$L = \tfrac{1}{2}(K + \dot{\varphi}_1\hat{A}_1D)E \cdot (K + \dot{\varphi}_1\hat{A}_1D) - \sigma^{\text{s}}(D). \qquad (5.4.25)$$

This shows that the generalized momentum associated with φ_1 is

$$p = \frac{\partial L}{\partial\dot{\varphi}_1} = \dot{\varphi}_1\hat{A}_1DE \cdot \hat{A}_1D + KE \cdot \hat{A}_1D. \qquad (5.4.26)$$

In analogy with (5.4.6) we set

$$J = \hat{A}_1DE \cdot \hat{A}_1D, \qquad (5.4.27)$$

and therefore

$$\dot{\varphi}_1 = J^{-1}(p - KE \cdot \hat{A}_1D). \qquad (5.4.28)$$

It is useful to rewrite the Lagrangian, using (5.4.26), in the form

$$L = -\tfrac{1}{2}J\dot{\varphi}_1^2 + p\dot{\varphi}_1 + \tfrac{1}{2}KE \cdot K - \sigma^{\text{s}}(D). \qquad (5.4.29)$$

Then the Routhian R defined by (5.4.5) becomes

$$R = \tfrac{1}{2}KE \cdot K - \tfrac{1}{2}J^{-1}(p - KE \cdot \hat{A}_1D)^2 - \sigma^{\text{s}}(D). \qquad (5.4.30)$$

If $R_{\text{R}} = I$, the deformation $(5.4.18)_1$ becomes the special case (5.3.1) considered in Section 5(iii). Therefore, the steady state considered there applies also in the general case here, provided we adjoint to it the values $\theta_1^0 = \theta_2^0 = \theta_3^0 = 0$. In particular, we obtain (5.4.4) once again, and the equilibrium value $[D^0] = \text{diag}(u_1^0, u_2^0, u_3^0)$ satisfies (5.3.10) with $\omega = \dot{\varphi}_1^0 = p/A$.

In order to examine the stability of this steady state, we must develop a quadratic approximation R' of the Routhian (5.4.30) near $\varphi_\alpha = 0$, $\theta_i = 0$, $D = D^0$. The calculations involved in forming this approximation are intricate. Rather than presenting them in detail, we restrict ourselves to recording the final result as well as some of the intermediate steps. Our principal concern is with the final form of R' and the implications it gives for a subsequent stability analysis.

From (4.1.12) we see that for α near 0,

$$Q_i(\alpha) = I + \alpha A_i + \tfrac{1}{2}\alpha^2 A_i^2 + \cdots . \tag{5.4.31}$$

Using this approximation repeatedly in (5.4.20), we find that

$$[\hat{A}_1] = \begin{bmatrix} 0 & -\varphi_2 & -\varphi_3 \\ \varphi_2 & 0 & -1 + \tfrac{1}{2}\varphi_2^2 + \tfrac{1}{2}\varphi_3^2 \\ \varphi_3 & 1 - \tfrac{1}{2}\varphi_2^2 - \tfrac{1}{2}\varphi_3^2 & 0 \end{bmatrix} + \cdots ,$$

$$[\hat{A}_2] = \begin{bmatrix} 0 & 0 & 1 - \tfrac{1}{2}\varphi_3^2 \\ 0 & 0 & -\varphi_3 \\ -1 + \tfrac{1}{2}\varphi_3^2 & \varphi_3 & 0 \end{bmatrix} + \cdots , \quad [\hat{A}_3] = \begin{bmatrix} 0 & -1 & 0 \\ 1 & 0 & 0 \\ 0 & 0 & 0 \end{bmatrix},$$

$$[\tilde{A}_1] = \begin{bmatrix} 0 & \theta_2 & \theta_3 \\ -\theta_2 & 0 & -1 + \tfrac{1}{2}\theta_2^2 + \tfrac{1}{2}\theta_3^2 \\ -\theta_3 & 1 - \tfrac{1}{2}\theta_2^2 - \tfrac{1}{2}\theta_3^2 & 0 \end{bmatrix} + \cdots ,$$

$$[\tilde{A}_2] = \begin{bmatrix} 0 & 0 & 1 - \tfrac{1}{2}\theta_3^2 \\ 0 & 0 & \theta_3 \\ -1 + \tfrac{1}{2}\theta_3^2 & -\theta_3 & 0 \end{bmatrix} + \cdots , \quad [\tilde{A}_3] = \begin{bmatrix} 0 & -1 & 0 \\ 1 & 0 & 0 \\ 0 & 0 & 0 \end{bmatrix}. \tag{5.4.32}$$

Similarly, from repeated use of (5.4.31) in (5.4.23), we obtain

$$[E] = \begin{bmatrix} E_1 + (E_3 - E_1)\theta_2^2 + (E_2 - E_1)\theta_3^2 \\ (E_1 - E_2)\theta_3 + (E_2 - E_3)\theta_1\theta_2 \\ (E_3 - E_1)\theta_2 + (E_3 - E_2)\theta_1\theta_3 \end{bmatrix}$$

$$\begin{bmatrix} (E_1 - E_2)\theta_3 + (E_2 - E_3)\theta_1\theta_2 & (E_3 - E_1)\theta_2 + (E_3 - E_2)\theta_1\theta_3 \\ E_2 + (E_3 - E_2)\theta_1^2 + (E_1 - E_2)\theta_3^2 & (E_2 - E_3)\theta_1 + (E_3 - E_1)\theta_2\theta_3 \\ (E_2 - E_3)\theta_1 + (E_3 - E_1)\theta_2\theta_3 & E_3 + (E_2 - E_3)\theta_1^2 + (E_1 - E_3)\theta_2^2 \end{bmatrix} + \cdots . \tag{5.4.33}$$

It is convenient to introduce perturbations of D in terms of a tensor δ satisfying

$$D = D^0(I + \delta) = (I + \delta)D^0. \tag{5.4.34}$$

Then the matrix $[\delta]$ is diagonal with components given explicitly by (5.3.21). From (5.4.32)$_1$ and (5.4.34), we obtain

$$[\hat{A}_1 D] =$$

$$\begin{bmatrix} 0 & -\varphi_2 - \varphi_2\delta_2 & -\varphi_3 - \varphi_3\delta_3 \\ \varphi_2 + \varphi_2\delta_1 & 0 & -1 - \delta_3 + \tfrac{1}{2}\varphi_2^2 + \tfrac{1}{2}\varphi_3^2 \\ \varphi_3 + \varphi_3\delta_1 & 1 + \delta_2 - \tfrac{1}{2}\varphi_2^2 - \tfrac{1}{2}\varphi_3^2 & 0 \end{bmatrix}[D^0] + \cdots . \tag{5.4.35}$$

Slightly more complex is the approximation of the term K given by (5.4.24). The building blocks are listed in (5.4.32) and (5.4.34). Placing these into (5.4.24) and retaining only linear and quadratic terms, we obtain

$$
[K] = \begin{bmatrix} \delta_1 & -\dot\phi_3 - \dot\theta_3\alpha & \dot\phi_2 + \dot\theta_2\beta \\ \dot\phi_3 + \dot\theta_3/\alpha & \delta_2 & -\dot\theta_1\gamma \\ -\dot\phi_2 - \dot\theta_2/\beta & \dot\theta_1/\gamma & \delta_3 \end{bmatrix} [D^0]
$$

$$
+ \begin{bmatrix} 0 & -\dot\phi_3\delta_2 + \theta_1\theta_2\alpha - \theta_3\delta_1\alpha & \dot\phi_2\delta_3 + \theta_1\theta_3\beta + \theta_2\delta_1\beta \\ \dot\phi_3\delta_1 - \theta_1\theta_2/\alpha + \theta_3\delta_2/\alpha & 0 & -\dot\phi_2\varphi_3 + \theta_2\theta_3\gamma - \theta_1\delta_2\gamma \\ -\dot\phi_2\delta_1 - \theta_1\theta_3/\beta - \theta_2\delta_3/\beta & \dot\phi_2\varphi_3 - \theta_2\theta_3/\gamma + \theta_1\delta_3/\gamma & 0 \end{bmatrix}
$$

$$
\cdot [D^0] + \cdots, \tag{5.4.36}
$$

where we have set

$$
\alpha = u_1^0/u_2^0, \qquad \beta = u_1^0/u_3^0, \qquad \gamma = u_2^0/u_3^0. \tag{5.4.37}
$$

In (5.4.33), (5.4.35), and (5.4.36), we have the three main approximations that are required to compute the terms $\frac{1}{2}KE \cdot K$, J as given by (5.4.27), and $KE \cdot \hat{A}_1 D$. It is now straightforward, though tedious, to form approximations of each of these terms and then to combine them to obtain the quadratic approximation R' of R. The result is

$$
R' = R_1 + R_2, \tag{5.4.38}
$$

where

$$
R_1 = \tfrac{1}{2}B\dot\phi_2^2 + \tfrac{1}{2}C\dot\phi_3^2 + \frac{p}{A}((A - B)\dot\phi_2\varphi_3 + C\dot\phi_3\varphi_2)
$$

$$
- \frac{p^2}{2A^2}(A + (A - C)\varphi_2^2 + (A - B)\varphi_3^2)
$$

$$
+ \tfrac{1}{2}(E_3^0\beta^2 + E_1^0/\beta^2)\dot\theta_2^2 + \tfrac{1}{2}(E_2^0\alpha^2 + E_1^0/\alpha^2)\dot\theta_3^2
$$

$$
- \frac{p}{A}((E_3^0\gamma + E_1^0/\alpha\beta)\dot\theta_2\theta_3 + (E_3^0\gamma - E_1^0/\alpha\beta)\dot\theta_3\theta_2)
$$

$$
- \frac{p^2}{2A^2}((E_3^0 - E_1^0/\beta^2)\theta_2^2 + (E_2^0 - E_1^0/\alpha^2)\theta_3^2)
$$

$$
+ (E_3^0\beta + E_1^0/\beta)\dot\phi_2\dot\theta_2 + (E_2^0\alpha + E_1^0/\alpha)\dot\phi_3\dot\theta_3 - \frac{p}{A}((E_3^0\beta + E_1^0/\beta)\dot\theta_2\varphi_3
$$

$$
- (E_2^0\alpha + E_1^0/\alpha)\dot\theta_3\varphi_2 + (E_3^0\beta - E_1^0/\beta)\dot\phi_3\theta_2 - (E_2^0\alpha - E_1^0/\alpha)\dot\phi_2\theta_3)
$$

$$
- \frac{p^2}{A^2}((E_3^0\beta - E_1^0/\beta)\varphi_2\theta_2 + (E_2^0\alpha - E_1^0/\alpha)\varphi_3\theta_3), \tag{5.4.39}
$$

$$R_2 = \tfrac{1}{2}E_i^0\dot{\delta}_i^2 + \frac{E_2^0 E_3^0}{2A}(\gamma - 1/\gamma)^2\dot{\theta}_1^2 + \frac{p}{A}\bigg((E_3^0\gamma + E_2^0/\gamma)\dot{\theta}_1$$

$$+ (E_3^0\gamma - E_2^0/\gamma)\theta_1(\dot{\delta}_2 - \dot{\delta}_3) + \frac{B - C}{A}(E_3^0\gamma + E_2^0/\gamma)\dot{\theta}_1(\delta_2 - \delta_3)\bigg)$$

$$- \frac{p^2}{2A^2}\bigg(\frac{3}{A}(E_2^0\delta_2 + E_3^0\delta_3)^2 - \frac{E_2^0 E_3^0}{A}(\delta_2 - \delta_3)^2$$

$$- \theta_1^2(E_3^0\gamma^2 + E_2^0/\gamma^2 - A)\bigg) - \sigma^0 - \tfrac{1}{2}\sigma_{ij}^0\delta_i\delta_j.$$

As before E_1^0, E_2^0, and E_3^0 are the diagonal elements of the matrix representation of the Euler tensor for the steady-state solution, and A, B, and C are given by (5.3.12).

From the perspective of small motions about steady spinning, there is a natural splitting of the Routhian into two effects. R_1 depends only on φ_2, φ_3, θ_2, θ_3, and their derivatives. The first few terms in R_1 are precisely the approximate Routhian for the case of a rigid body. Indeed, if we specialize the analysis here to small motions in which $\theta_2 \equiv 0$ and $\theta_3 \equiv 0$, the conventional results on the stability of a spinning rigid body carry over to a pseudo-rigid body. This is essentially the case that we treated in Section 5(iii), though from a formulation in terms of the components of angular velocity.

The term R_2 depends only on δ_1, δ_2, δ_3, θ_1, and their derivatives. Thus, an analysis of stability gives rise to two decoupled effects. From R_1 we obtain a system in which the angles of rotation φ_2 and φ_3 in planes containing the axis of rotation interact with the two shears, represented by θ_2 and θ_3, in these same planes. Instabilities arising from this system tend to "tip" the axis of rotation away from its location in the steady state, both by rotation and shearing of the body. We may expect for this system instabilities generalizing those exhibited by a rigid body. From R_2 we get a system linking the principal stretches to the shear, represented by θ_1, in the plane perpendicular to the steady-state axis of rotation. If we take as a reasonable guide the special analysis for a St. Vénant–Kirchhoff material presented in Section 5(iii), this coupled system may indeed exhibit stable motions.

CHAPTER 6

Approximations for Almost-rigid Bodies

If for all motions in a certain class the deformations of a body are small enough to be ignored, then relative to this class we say the body is rigid. The assumption of rigidity is an idealization that has given rise to a large body of work in mechanics. Nevertheless, there are also important problems in which the effects of deformation, while still quite small, cannot be ignored. For motions of this type we say the body is nearly rigid. Approximations for nearly rigid bodies have technological implications both old and new. Historically, the gyroscopic motion of the earth has always fascinated and challenged astronomers and geophysicists. This motion exhibits special complexities owing to the fact that the earth is not exactly rigid. An indication of the scope of work in this area can be found in the treatise of LAMBECK [1980]. More currently, problems arise in space technology where the control of orbital motions of space vehicles and satellites is paramount. Complexities arise here from the flexibility of the satellite and accompanying elastic oscillations and rotational perturbations. The analysis of such problems involves rather ingenious modeling and large-scale numerical computations. The work of WILLIAMS [1976] provides an indication of the state of the art in the field.

In this chapter, we use the model of a pseudo-rigid elastic body to analyze the gyroscopic motion of flexible bodies. In particular, we consider the force-free motion of an axially symmetric body with a fixed point. At least to a limited extent we have addressed such problems previously. In Section 4(ii), we found the deformation in a body that is spinning freely about its axis of symmetry; in Section 4(i), we analyzed steady precession for a body that is constrained to deform rigidly. Much more difficult is the problem of steady precession in the absence of applied loadings, and this is our concern here. For general pseudo-rigid bodies, nonlinear interactions that occur between rotation and deformation make an analysis of this problem prohibitive. Consequently, we are led to consider some simplifying process of linearization. Specifically, our objective is *to extract from the equations for pseudo-rigid bodies an approximate solution that gives prominence to the rigid motion but at the same time retains the elastic effects in the sense of linear elasticity.* As we shall see later, the flexibility of the body contributes to the deformation through a perturbation of a rigid motion. Since the body is nearly but not completely rigid, we are led then to the study of *almost-rigid* bodies.

(i) The Problem of Transition to Rigidity

In terms of the polar decomposition $F = RU$, we wish to consider approximate solutions of the equations of motion

$$\ddot{F}E_R = M_R - \Sigma_R \tag{6.1.1}$$

in which U is close to the identity I. Setting

$$U = I + S, \qquad \Omega = R^T \dot{R} \tag{6.1.2}$$

(cf. (4.3.6) and (5.2.21)), we note that

$$\dot{F} = R(\dot{S} + \Omega(I + S)),$$
$$\ddot{F} = R(\ddot{S} + 2\Omega\dot{S} + (\dot{\Omega} + \Omega^2)(I + S)), \tag{6.1.3}$$

and so the differential equation (6.1.1) can be written in the form

$$(\ddot{S} + 2\Omega\dot{S} + (\dot{\Omega} + \Omega^2)(I + S))E_R = -R^T\Sigma_R + \overline{M}, \tag{6.1.4}$$

where $\overline{M} = R^T M_R$. Since the tensor S is near 0, we select a response function that is linear in S. For simplicity, we consider only an isotropic material, and then $(2.7.2)_1$, $(2.7.13)_1$, and $(2.7.29)_3$ provide the classical linear form

$$\Sigma_R = \mathfrak{S}_R(F)$$
$$= R\mathfrak{S}_u(I + S)$$
$$= R(\lambda(\mathrm{tr}\, S)I + 2\mu S), \tag{6.1.5}$$

where λ and μ are Lamé moduli satisfying (2.7.28).

As μ and λ become large, the principal force-moments in such a material become unbounded unless the strain tensor S simultaneously approaches 0. This suggests that for a transition to rigidity we should replace λ and μ by λ/ε^2 and μ/ε^2, respectively, and consider the limit $\varepsilon \to 0$. Simple examples of the effects of this limit show[1] that, in order for a rigid state to arise as $\varepsilon \to 0$, $S(0)$ should be of the order ε^2 and $\dot{S}(0)$ should be of order ε. Thus, we restate the basic problem in the form

---

[1] Such examples can be constructed from scalar analogues of the equation of motion (6.1.1).

$$(\ddot{S} + 2\Omega\dot{S} + (\dot{\Omega} + \Omega^2)(I + S))E_{\mathrm{R}} = -\frac{\lambda}{\varepsilon^2}(\mathrm{tr}\ S)I - \frac{2\mu}{\varepsilon^2}S + \overline{M},$$

$$\Omega(0) = \Omega_0, \qquad S(0) = \varepsilon^2 K, \qquad \dot{S}(0) = \varepsilon L, \tag{6.1.6}$$

where Ω_0 is skew, and K and L are symmetric. We assume that Ω_0, K, L, and $\overline{M}(t)$ are independent of ε. As a final reduction we write

$$\Omega = \Omega^\varepsilon, \qquad S = \varepsilon^2 S^\varepsilon. \tag{6.1.7}$$

Then (6.1.6) becomes

$$(\varepsilon^2 \ddot{S}^\varepsilon + 2\varepsilon^2 \Omega^\varepsilon \dot{S}^\varepsilon + (\dot{\Omega}^\varepsilon + \Omega^{\varepsilon 2})(I + \varepsilon^2 S^\varepsilon))E_{\mathrm{R}}$$

$$= -\lambda(\mathrm{tr}\ S^\varepsilon)I - 2\mu S^\varepsilon + \overline{M}, \tag{6.1.8}$$

$$\Omega^\varepsilon(0) = \Omega_0, \qquad S^\varepsilon(0) = K, \qquad \dot{S}^\varepsilon(0) = L/\varepsilon.$$

The analysis of (6.1.8) is intricate and somewhat long. For clarity, we present the main results now in the form of two theorems, deferring proofs until later sections. The first, the Spectral Theorem, gives the general solution of a nonhomogeneous second-order linear matrix differential equation that arises at several stages of our analysis. A second result, the Transition Theorem, asserts a two-time-scale asymptotic series representation for the motion of bodies that are nearly rigid.

Spectral Theorem. *Let* $\mathbb{L}$ *and* $\mathbb{M}$ *be the operators*

$$\mathbb{L}A = \lambda(\mathrm{tr}\ A)I + 2\mu\ \mathrm{sym}(E_{\mathrm{R}}^{1/2}AE_{\mathrm{R}}^{-1/2}),$$

$$\mathbb{M}A = E_{\mathrm{R}}^{1/2}AE_{\mathrm{R}}^{1/2}, \tag{6.1.9}$$

on the space of symmetric tensors. Then the eigenvalue problem

$$(\mathbb{L} - \eta\mathbb{M})A = 0 \tag{6.1.10}$$

has six positive real eigenvalues $\sigma_1^2, \ldots, \sigma_6^2$. *The associated eigenvectors are symmetric tensors* $A_1, \ldots, A_6$ *that are orthonormal in the sense*

$$A_i \cdot E_{\mathrm{R}}^{1/2} A_j E_{\mathrm{R}}^{1/2} = \delta_{ij}. \tag{6.1.11}$$

In terms of these, the general solution of the matrix differential equation

$$\frac{d^2}{d\tau^2}S + \lambda(\mathrm{tr}\ S)E_{\mathrm{R}}^{-1} + 2\mu\ \mathrm{sym}(SE_{\mathrm{R}}^{-1})$$

$$= \sum_{i=1}^{6}(C_i \sin \sigma_i\tau + D_i \cos \sigma_i\tau) + G, \tag{6.1.12}$$

where C_i, D_i, *and* G *are constant symmetric tensors, is*

$$S(\tau) = \sum_{i=1}^{6} (\alpha_i \cos \sigma_i \tau + \beta_i \sin \sigma_i \tau) A_i$$

$$+ \sum_{i=1}^{6} \sum_{j \in I_i} \tfrac{1}{2}(\sigma_i)^{-1}(A_j \cdot E_R^{1/2} D_i E_R^{1/2} \tau \sin \sigma_i \tau$$

$$- A_j \cdot E_R^{1/2} C_i E_R^{1/2} \tau \cos \sigma_i \tau) A_j$$

$$+ \sum_{i=1}^{6} \sum_{j \notin I_i} (\sigma_j^2 - \sigma_i^2)^{-1}(A_j \cdot E_R^{1/2} C_i E_R^{1/2} \sin \sigma_i \tau$$

$$+ A_j \cdot E_R^{1/2} D_i E_R^{1/2} \cos \sigma_i \tau) A_j + S^0. \tag{6.1.13}$$

S^0 is the unique constant symmetric tensor satisfying

$$\lambda(\operatorname{tr} S^0) E_R^{-1} + 2\mu \operatorname{sym}(S^0 E_R^{-1}) = G, \tag{6.1.14}$$

α_i and β_i, $i = 1, \ldots, 6$, are constants of integration, and I_i is the set of indices $\{j \,|\, \sigma_j^2 = \sigma_i^2\}$.

Transition Theorem. *The solution of* (6.1.8) *has a formal asymptotic expansion*

$$\Omega^\varepsilon(t) = \bar{\Omega}^{(0)}(t) + \varepsilon \bar{\Omega}^{(1)}(t) + \varepsilon \hat{\Omega}^{(1)}(t, t/\varepsilon) + O(\varepsilon^2),$$
$$S^\varepsilon(t) = \bar{S}^{(0)}(t) + \hat{S}^{(0)}(t, t/\varepsilon) + O(\varepsilon), \tag{6.1.15}$$

that is valid uniformly in t as $\varepsilon \to 0$. $\bar{\Omega}^{(0)}$ and $\bar{\Omega}^{(1)}$ are the solutions of the initial-value problems

$$\operatorname{sk}((\dot{\bar{\Omega}}^{(0)} + \bar{\Omega}^{(0)2}) E_R) = \operatorname{sk} \bar{M},$$
$$\bar{\Omega}^{(0)}(0) = \Omega_0, \tag{6.1.16}$$

and

$$\operatorname{sk}((\dot{\bar{\Omega}}^{(1)} + \bar{\Omega}^{(0)}\bar{\Omega}^{(1)} + \bar{\Omega}^{(1)}\bar{\Omega}^{(0)}) E_R) = 0,$$
$$\bar{\Omega}^{(1)}(0) = -2\mu \sum_{i=1}^{6} (\sigma_i^2)^{-1} L \cdot E_R^{1/2} A_i E_R^{1/2} \operatorname{sk}(A_i E_R^{-1}), \tag{6.1.17}$$

where σ_i and A_i are the eigenvalues and eigenvectors from the Spectral Theorem. $\bar{S}^{(0)}$ is the unique symmetric tensor satisfying

$$\lambda(\operatorname{tr} \bar{S}^{(0)}) I + 2\mu \bar{S}^{(0)} = \operatorname{sym}(\bar{M} - (\dot{\bar{\Omega}}^{(0)} + \bar{\Omega}^{(0)2}) E_R). \tag{6.1.18}$$

If

$$p_{ji} = \operatorname{sym}\left(\bar{\Omega}^{(0)} A_i + \frac{2\mu}{\sigma_i^2} \bar{\Omega}^{(0)} \operatorname{sk}(A_i E_R^{-1})\right) \cdot E_R^{1/2} A_j E_R^{1/2}, \tag{6.1.19}$$

and if α_i and β_i, $i = 1, \ldots, 6$, are the unique solutions of the initial-value problems

$$\dot{\alpha}_i + \sum_{j \in I_i} p_{ij}(t)\alpha_j = 0, \qquad \alpha_i(0) = (K - \bar{S}^{(0)}(0)) \cdot E_R^{1/2} A_i E_R^{1/2},$$

$$\dot{\beta}_i + \sum_{j \in I_i} p_{ij}(t)\beta_j = 0, \qquad \beta_i(0) = (\sigma_i)^{-1} L \cdot E_R^{1/2} A_i E_R^{1/2},$$

(6.1.20)

then

$$\mathbf{\Omega}^{(1)}(t, \tau) = -2\mu \sum_{i=1}^{6} (\sigma_i)^{-1} \, \mathrm{sk}(A_i E_R^{-1})(\alpha_i(t) \sin \sigma_i \tau - \beta_i(t) \cos \sigma_i \tau),$$

(6.1.21)

$$\mathbf{S}^{(0)}(t, \tau) = \sum_{i=1}^{6} A_i(\alpha_i(t) \cos \sigma_i \tau + \beta_i(t) \sin \sigma_i \tau).$$

The initial-value problem (6.1.16) governs the leading approximation of $\mathbf{\Omega}^\varepsilon$ and presents a problem in classical rigid-body mechanics (cf. (4.1.1) and $(6.1.2)_2$). The leading term in the approximation of S^ε, as determined by (6.1.18), is precisely the strain found by balancing force-moments in the body against the portion of the external force-moment $\overline{M}$ that is not accounted for by the rigid motion. The next perturbation, found by solving (6.1.17), provides a correction of order ε to the leading rotation term and arises as a consequence of initial strain and strain rates. Subsequent steps in the perturbation follow a similar pattern. In Sections 6(iv) and 6(v), we specialize these results to axially symmetric bodies and examine the form of the asymptotic expansion when the rigid-body motion corresponds to precession.

(ii) Derivation of the Asymptotic Expansion

The pure stretch solution obtained in Section 4(iii) (cf. (4.3.11)) shows that in a transition to a rigid state we may expect high-frequency oscillations, typically varying on a fast time scale given by

$$\tau = t/\varepsilon. \tag{6.2.1}$$

At the same time, we may expect the rigid-body motion to proceed on the normal time scale given by t. This suggests a two-time method for the perturbation analysis.[2] In this approach, we consider expansions

$$\mathbf{\Omega}^\varepsilon(t) = \sum_{k=0}^{\infty} \varepsilon^k \mathbf{\Omega}^{(k)}(t, t/\varepsilon),$$

$$\mathbf{S}^\varepsilon(t) = \sum_{k=0}^{\infty} \varepsilon^k \mathbf{S}^{(k)}(t, t/\varepsilon),$$

(6.2.2)

where the successive terms must satisfy the *secularity condition*:

$$\mathbf{\Omega}^{(k)}(t, \tau), \ \mathbf{S}^{(k)}(t, \tau), \text{ and } \mathbf{S}_\tau^{(k)}(t, \tau) \text{ are uniformly bounded}$$
in τ for $0 < \tau < \infty$.

(6.2.3)

---

[2] Cf. GREENBERG [1978, Chap. 25].

Here and henceforth partial derivatives with respect to t and τ are denoted by subscripts and total derivatives with respect to t are denoted by an over dot. For example, from (6.2.2) we find that

$$\dot{\Omega}^{\varepsilon} = \sum_{k=0}^{\infty} \varepsilon^k (\Omega_t^{(k)} + \Omega_\tau^{(k)} \varepsilon^{-1})$$

$$= \Omega_\tau^{(0)} \varepsilon^{-1} + \sum_{k=0}^{\infty} \varepsilon^k (\Omega_t^{(k)} + \Omega_\tau^{(k+1)}). \tag{6.2.4}$$

Similarly,

$$\dot{S}^{\varepsilon} = S_\tau^{(0)} \varepsilon^{-1} + \sum_{k=0}^{\infty} \varepsilon^k (S_t^{(k)} + S_\tau^{(k+1)}),$$

$$\ddot{S}^{\varepsilon} = S_{\tau\tau}^{(0)} \varepsilon^{-2} + (2S_{t\tau}^{(0)} + S_{\tau\tau}^{(1)}) \varepsilon^{-1} \tag{6.2.5}$$

$$+ \sum_{k=0}^{\infty} \varepsilon^k (S_{tt}^{(k)} + 2S_{t\tau}^{(k+1)} + S_{\tau\tau}^{(k+2)}).$$

When these expansions are placed in (6.1.8), an iterative system arises from separating out the coefficients of similar powers of ε. The calculation is tedious but straightforward, so we simply list the results. For the terms of lowest order in the expansions of Ω^{ε} and S^{ε}, we obtain

Order ε^{-1}: $\quad \Omega_\tau^{(0)} E_R = 0, \qquad S_\tau^{(0)}(0,0) = L,$ $\hfill$ (6.2.6)

Order 1: $\quad (S_{\tau\tau}^{(0)} + \Omega_t^{(0)} + \Omega_\tau^{(1)} + \Omega^{(0)2}) E_R = -\lambda(\operatorname{tr} S^{(0)}) I - 2\mu S^{(0)} + \overline{M},$

$$\Omega^{(0)}(0,0) = \Omega_0, \qquad S^{(0)}(0,0) = K, \tag{6.2.7}$$

$$S_t^{(0)}(0,0) + S_\tau^{(1)}(0,0) = 0,$$

Order ε: $\quad (S_{\tau\tau}^{(1)} + 2S_{t\tau}^{(0)} + 2\Omega^{(0)} S_\tau^{(0)} + \Omega_\tau^{(2)} + \Omega_t^{(1)} + \Omega_t^{(0)} S^{(0)}$

$$+ \Omega^{(0)}\Omega^{(1)} + \Omega^{(1)}\Omega^{(0)}) E_R = -\lambda(\operatorname{tr} S^{(1)}) I - 2\mu S^{(1)},$$

$$\Omega^{(1)}(0,0) = 0, \qquad S^{(1)}(0,0) = 0, \tag{6.2.8}$$

$$S_t^{(1)}(0,0) + S_\tau^{(2)}(0,0) = 0.$$

By $(6.2.6)_1$ the coefficient $\Omega^{(0)}$ is a function of t alone. We denote it by $\overline{\Omega}^{(0)}$, henceforth using an over bar to signify a function of t alone. In particular, we write $\dot{\overline{\Omega}}^{(0)}$ in place of $\Omega_t^{(0)}$.

Consider $(6.2.7)_1$ next. Multiplying by E_R^{-1} and separating into symmetric and skew parts, we obtain the two equations

$$S_{\tau\tau}^{(0)} + \lambda(\operatorname{tr} S^{(0)}) E_R^{-1} + 2\mu \operatorname{sym}(S^{(0)} E_R^{-1}) = \operatorname{sym}(\overline{H} E_R^{-1}),$$

$$\Omega_\tau^{(1)} = \operatorname{sk}(\overline{H} E_R^{-1}) - 2\mu \operatorname{sk}(S^{(0)} E_R^{-1}), \tag{6.2.9}$$

where

$$\overline{H} = \overline{M} - (\dot{\overline{\Omega}}^{(0)} + \overline{\Omega}^{(0)2}) E_R. \tag{6.2.10}$$

Note that $\bar{H}$ is a function of t alone. By the Spectral Theorem the general solution of $(6.2.9)_1$ is

$$S^{(0)}(t, \tau) = \sum_{i=1}^{6} A_i(\alpha_i(t) \cos \sigma_i \tau + \beta_i(t) \sin \sigma_i \tau) + \bar{S}^{(0)}(t), \qquad (6.2.11)$$

where $\bar{S}^{(0)}$ is the unique symmetric tensor satisfying

$$\lambda(\operatorname{tr} \bar{S}^{(0)}) E_R^{-1} + 2\mu \operatorname{sym}(\bar{S}^{(0)} E_R^{-1}) = \operatorname{sym}(\bar{H} E_R^{-1}). \qquad (6.2.12)$$

Placing this solution for $S^{(0)}$ into $(6.2.9)_2$ and simplifying, we obtain

$$\Omega_\tau^{(1)} = \operatorname{sk}((\bar{H} - 2\mu \bar{S}^{(0)}) E_R^{-1})$$

$$- 2\mu \sum_{i=1}^{6} \operatorname{sk}(A_i E_R^{-1})(\alpha_i(t) \cos \sigma_i \tau + \beta_i(t) \sin \sigma_i \tau). \qquad (6.2.13)$$

The first term on the right-hand side must vanish. Otherwise, $\Omega^{(1)}$ will have a term that is linear in τ and hence will violate the secularity condition (6.2.3). Thus,

$$\operatorname{sk}((\bar{H} - 2\mu \bar{S}^{(0)}) E_R^{-1}) = \mathbf{0}. \qquad (6.2.14)$$

Now we can integrate (6.2.13) with respect to τ to obtain

$$\Omega^{(1)}(t, \tau) = -2\mu \sum_{i=1}^{6} \sigma_i^{-1} \operatorname{sk}(A_i E_R^{-1})(\alpha_i(t) \sin \sigma_i \tau$$

$$- \beta_i(t) \cos \sigma_i \tau) + \bar{\Omega}^{(1)}(t) \qquad (6.2.15)$$

for some function $\bar{\Omega}^{(1)}$ of t alone. These results are (6.1.15) and (6.1.21) of the Transition Theorem. If we rewrite (6.2.12) in the form

$$\operatorname{sym}((\bar{H} - 2\mu \bar{S}^{(0)}) E_R^{-1}) = \lambda(\operatorname{tr} \bar{S}^{(0)}) E_R^{-1} \qquad (6.2.16)$$

and recall (6.2.14), we see that

$$(\bar{H} - 2\mu \bar{S}^{(0)}) E_R^{-1} = \lambda(\operatorname{tr} \bar{S}^{(0)}) E_R^{-1}. \qquad (6.2.17)$$

Canceling E_R^{-1} and separating into symmetric and skew parts, we obtain $(6.1.16)_1$ and (6.1.18) of the Transition Theorem. Also, $(6.1.16)_2$ is simply $(6.2.7)_2$.

In order to determine $\bar{\Omega}^{(1)}$ and the scalars $\alpha_i(t)$ and $\beta_i(t)$, we must analyze in a similar fashion the next equation in the iterative system, namely, $(6.2.8)_1$. Multiplying that equation by E_R^{-1} and separating into symmetric and skew parts, we obtain the two problems

$$S_{\tau\tau}^{(1)} + \lambda(\operatorname{tr} S^{(1)}) E_R^{-1} + 2\mu \operatorname{sym}(S^{(1)} E_R^{-1})$$

$$= -2S_{t\tau}^{(0)} - 2 \operatorname{sym}(\bar{\Omega}^{(0)} S_\tau^{(0)}) - \bar{\Omega}^{(0)} \Omega^{(1)} - \Omega^{(1)} \bar{\Omega}^{(0)}, \qquad (6.2.18)$$

$$\Omega_\tau^{(2)} = -2 \operatorname{sk}(\bar{\Omega}^{(0)} S_\tau^{(0)}) - \Omega_t^{(1)} - 2\mu \operatorname{sk}(S^{(1)} E_R^{-1}).$$

Into the first equation we substitute the expressions (6.2.11) and (6.2.15) for $S^{(0)}$ and $\Omega^{(1)}$, respectively, and then simplify the right-hand side. The result is

$$S_{\tau\tau}^{(1)} + \lambda(\operatorname{tr} S^{(1)})E_R^{-1} + 2\mu \operatorname{sym}(S^{(1)} E_R^{-1})$$

$$= \sum_{i=1}^{6} \sin \sigma_i \tau \{2\dot{\alpha}_i \sigma_i A_i + 2\alpha_i \sigma_i \operatorname{sym}(\bar{\Omega}^{(0)} A_i)$$

$$+ \alpha_i(4\mu/\sigma_i) \operatorname{sym}(\bar{\Omega}^{(0)} \operatorname{sk}(A_i E_R^{-1}))\}$$

$$+ \sum_{i=1}^{6} \cos \sigma_i \tau \{2\dot{\beta}_i \sigma_i A_i + 2\beta_i \sigma_i \operatorname{sym}(\bar{\Omega}^{(0)} A_i)$$

$$+ \beta_i(4\mu/\sigma_i) \operatorname{sym}(\bar{\Omega}^{(0)} \operatorname{sk}(A_i E_R^{-1}))\} - \{\bar{\Omega}^{(0)}\bar{\Omega}^{(1)} + \bar{\Omega}^{(1)}\bar{\Omega}^{(0)}\}. \quad (6.2.19)$$

We have written this so as to emphasize that each term within brace brackets is a function of t alone. The general solution of this equation contains certain secular terms that must be eliminated. By the Spectral Theorem, the general solution has the form (6.1.13). With $I_j = \{i|\sigma_i^2 = \sigma_j^2\}$, $j = 1, \ldots, 6$, the terms there involving $\tau \sin \sigma_i \tau$ and $\tau \cos \sigma_i \tau$ will vanish if and only if

$$E_R^{1/2} A_j E_R^{1/2} \cdot \sum_{i \in I_j} C_i = 0, \qquad E_R^{1/2} A_j E_R^{1/2} \cdot \sum_{i \in I_j} D_i = 0, \qquad j = 1, \ldots, 6. \quad (6.2.20)$$

For our application C_i and D_i are given by the quantities in the first two brace brackets in (6.2.19). Placing these into (6.2.20) and simplifying by use of the orthogonality relation (6.1.11), we obtain (6.1.19), (6.1.20)$_1$, and (6.1.20)$_3$ of the Transition Theorem.

We must carry the analysis of (6.2.18) one step further. By the Spectral Theorem and our elimination of secular terms, we see that $S^{(1)}$ has the form

$$S^{(1)}(t, \tau) = \sum_{i=1}^{6} (U_i(t) \sin \sigma_i \tau + V_i(t) \cos \sigma_i \tau) + \bar{S}^{(1)}(t), \quad (6.2.21)$$

where $\bar{S}^{(1)}$ is the unique symmetric tensor-valued function of t alone that satisfies

$$\lambda(\operatorname{tr} \bar{S}^{(1)})E_R^{-1} + 2\mu \operatorname{sym}(\bar{S}^{(1)} E_R^{-1}) = -(\bar{\Omega}^{(0)}\bar{\Omega}^{(1)} + \bar{\Omega}^{(1)}\bar{\Omega}^{(0)}). \quad (6.2.22)$$

Placing this solution for $S^{(1)}$ into (6.2.18)$_2$ and simplifying, we obtain the differential equation

$$\Omega_\tau^{(2)} = -\dot{\bar{\Omega}}^{(1)} - 2\mu \operatorname{sk}(\bar{S}^{(1)} E_R^{-1}) + \text{terms in } \sin \sigma_i \tau \text{ and } \cos \sigma_i \tau. \quad (6.2.23)$$

We see from this that

$$-\dot{\bar{\Omega}}^{(1)} - 2\mu \operatorname{sk}(\bar{S}^{(1)} E_R^{-1}) = 0; \quad (6.2.24)$$

otherwise, $\Omega^{(2)}$ would contain a term proportional to τ. Subtracting (6.2.24) from (6.2.22), we obtain

$$\lambda(\text{tr } \bar{\boldsymbol{S}}^{(1)})\boldsymbol{E}_\text{R}^{-1} + 2\mu \bar{\boldsymbol{S}}^{(1)}\boldsymbol{E}_\text{R}^{-1} + \overset{\bullet}{\bar{\boldsymbol{\Omega}}}{}^{(1)} = -(\bar{\boldsymbol{\Omega}}^{(0)}\bar{\boldsymbol{\Omega}}^{(1)} + \bar{\boldsymbol{\Omega}}^{(1)}\bar{\boldsymbol{\Omega}}^{(0)}). \quad (6.2.25)$$

Multiplying by $\boldsymbol{E}_\text{R}$ and taking the skew part, we obtain $(6.1.17)_1$ of the Transition Theorem.

Finally, we need the initial conditions $(6.1.17)_2$, $(6.1.20)_2$, and $(6.1.20)_4$. Using (6.2.11), we see that the initial conditions $(6.2.7)_3$ and $(6.2.6)_2$ become

$$\sum_{i=1}^{6} A_i \alpha_i(0) + \bar{\boldsymbol{S}}^{(0)}(0) = \boldsymbol{K},$$

$$\sum_{i=1}^{6} A_i \sigma_i \beta_i(0) = \boldsymbol{L}. \quad (6.2.26)$$

By the orthogonality relation (6.1.11) these are equivalent to $(6.1.20)_2$ and $(6.1.20)_4$. Finally, using (6.2.15) we find that the initial condition $(6.2.8)_2$ becomes

$$0 = 2\mu \sum_{i=1}^{6} \sigma_i^{-1} \text{ sk}(A_i \boldsymbol{E}_\text{R}^{-1})\beta_i(0) + \bar{\boldsymbol{\Omega}}^{(1)}(0). \quad (6.2.27)$$

By $(6.1.20)_4$ this is equivalent to $(6.1.17)_2$. The proof of the Transition Theorem is complete.

(iii) Proof of the Spectral Theorem

As a preliminary to the proof of the Spectral Theorem, we establish the following lemma.

Lemma. *The operators* $\mathbb{L}$ *and* $\mathbb{M}$ *defined on the space of symmetric tensors by* (6.1.9) *are both positive-definite and symmetric.*

Proof. For any symmetric tensors A and B,

$$\boldsymbol{B} \cdot \mathbb{M}\boldsymbol{A} = \boldsymbol{B} \cdot \boldsymbol{E}_\text{R}^{1/2} \boldsymbol{A} \boldsymbol{E}_\text{R}^{1/2} = \boldsymbol{E}_\text{R}^{1/2} \boldsymbol{B} \boldsymbol{E}_\text{R}^{1/2} \cdot \boldsymbol{A} = \boldsymbol{A} \cdot \mathbb{M}\boldsymbol{B} \quad (6.3.1)$$

and

$$\begin{aligned}
\boldsymbol{B} \cdot \mathbb{L}\boldsymbol{A} &= \lambda(\text{tr } \boldsymbol{A})\boldsymbol{B} \cdot \boldsymbol{I} + \mu(\boldsymbol{B} \cdot \boldsymbol{E}_\text{R}^{1/2} \boldsymbol{A} \boldsymbol{E}_\text{R}^{-1/2} + \boldsymbol{B} \cdot \boldsymbol{E}_\text{R}^{-1/2} \boldsymbol{A} \boldsymbol{E}_\text{R}^{1/2}) \\
&= \lambda(\text{tr } \boldsymbol{A})(\text{tr } \boldsymbol{B}) + \mu(\boldsymbol{E}_\text{R}^{1/2} \boldsymbol{B} \boldsymbol{E}_\text{R}^{-1/2} \cdot \boldsymbol{A} + \boldsymbol{E}_\text{R}^{-1/2} \boldsymbol{B} \boldsymbol{E}_\text{R}^{1/2} \cdot \boldsymbol{A}) \\
&= \boldsymbol{A} \cdot \mathbb{L}\boldsymbol{B}. \quad (6.3.2)
\end{aligned}$$

Thus, $\mathbb{M}$ and $\mathbb{L}$ are symmetric operators. Also,

$$\boldsymbol{A} \cdot \mathbb{M}\boldsymbol{A} = \boldsymbol{A} \cdot \boldsymbol{E}_\text{R}^{1/2} \boldsymbol{A} \boldsymbol{E}_\text{R}^{1/2} = \|\boldsymbol{E}_\text{R}^{1/4} \boldsymbol{A} \boldsymbol{E}_\text{R}^{1/4}\|^2, \quad (6.3.3)$$

and this shows that $\mathbb{M}$ is positive-definite. Finally, we note that

$$A \cdot \mathbb{L}A = \lambda(\text{tr } A)^2 + \mu A \cdot E_R^{1/2} A E_R^{-1/2} + \mu A \cdot E_R^{-1/2} A E_R^{1/2}$$

$$= \frac{3\lambda + 2\mu}{3}(\text{tr } A)^2 + \mu \hat{A} \cdot E_R^{1/2} \hat{A} E_R^{-1/2} + \mu \hat{A} \cdot E_R^{-1/2} \hat{A} E_R^{1/2}$$

$$= \frac{3\lambda + 2\mu}{3}(\text{tr } A)^2 + \mu(E_R^{-1/4} \hat{A} E_R^{-1/4}) \cdot E_R(E_R^{-1/4} \hat{A} E_R^{-1/4})$$

$$+ \mu(E_R^{1/4} \hat{A} E_R^{1/4}) \cdot E_R^{-1}(E_R^{1/4} \hat{A} E_R^{1/4}), \tag{6.3.4}$$

where

$$\hat{A} = A - \tfrac{1}{3}(\text{tr } A)I. \tag{6.3.5}$$

Given any symmetric tensor C with the ith row denoted by c_i, we see that

$$C \cdot E_R C = \text{tr}(CE_R C) = \sum_{i=1}^{3} c_i \cdot E_R c_i. \tag{6.3.6}$$

Since E_R is positive-definite, we conclude that $C \cdot E_R C \geq 0$ with equality only if $C = 0$. This fact and the inequalities (2.7.28) show that the three terms on the right-hand side of $(6.3.4)_3$ are all nonnegative and vanish only if $\hat{A} = 0$ and $\text{tr } A = 0$, that is, only if $A = 0$. Hence, $\mathbb{L}$ is positive-definite, and this completes the proof.

The properties of the eigenvalue problem (6.1.10) stated in the Spectral Theorem follow directly from the above lemma and standard results in linear algebra.

Let us turn now to (6.1.12). The theory of linear differential equations tells us that the homogeneous problem

$$\frac{d^2}{d\tau^2}S + \lambda(\text{tr } S)E_R^{-1} + 2\mu \, \text{sym}(SE_R^{-1}) = 0 \tag{6.3.7}$$

has a complete set of linearly independent solutions of the form

$$S = e^{\xi\tau}A. \tag{6.3.8}$$

Placing this in (6.3.7), we find that ξ and A must satisfy

$$\xi^2 A + \lambda(\text{tr } A)E_R^{-1} + 2\mu \, \text{sym}(AE_R^{-1}) = 0. \tag{6.3.9}$$

Multiplying on both the left and the right by $E_R^{1/2}$ and setting $\eta = -\xi^2$, we obtain the eigenvalue problem (6.1.10). It follows that the general solution of (6.3.7) has the form

$$S = \sum_{i=1}^{6} A_i(\alpha_i \cos \sigma_i \tau + \beta_i \sin \sigma_i \tau), \tag{6.3.10}$$

where α_i and β_i, $i = 1, \ldots, 6$, are to be determined from initial conditions.
Next we examine the nonhomogeneous case. Consider first the problem

$$\frac{d^2}{d\tau^2}S + \lambda(\text{tr } S)E_R^{-1} + 2\mu \text{ sym}(SE_R^{-1}) = G, \qquad (6.3.11)$$

where G is a constant symmetric tensor, and look for a particular solution that is constant, say S^0. Thus,

$$\lambda(\text{tr } S^0)E_R^{-1} + 2\mu \text{ sym}(S^0 E_R^{-1}) = G. \qquad (6.3.12)$$

Equivalently, there is a skew tensor T such that

$$2\mu S^0 E_R^{-1} = G - \lambda(\text{tr } S^0)E_R^{-1} + T. \qquad (6.3.13)$$

This equation is easily solved for S^0, the result being

$$S^0 = \frac{1}{2\mu}\left((G - T)E_R - \frac{\lambda}{3\lambda + 2\mu}(\text{tr}((G - T)E_R))I\right). \qquad (6.3.14)$$

Since S^0 is symmetric, T must be selected such that

$$\text{sk}((G - T)E_R) = 0. \qquad (6.3.15)$$

This can be rewritten as

$$TE_R + E_R T = GE_R - E_R G \qquad (6.3.16)$$

or as

$$\mathbb{P}T = GE_R - E_R G, \qquad (6.3.17)$$

where $\mathbb{P}$ is the following operator on the space of skew tensors:

$$\mathbb{P}T = TE_R + E_R T. \qquad (6.3.18)$$

By mimicking the proof of the preceding lemma, we can show that $\mathbb{P}$ is positive-definite and symmetric. Hence, $\mathbb{P}$ is invertible and so there is a unique skew tensor T satisfying (6.3.15).

Finally, we consider the nonhomogeneous equation

$$\frac{d^2}{d\tau^2}S + \lambda(\text{tr } S)E_R^{-1} + 2\mu \text{ sym}(SE_R^{-1}) = C \sin \sigma\tau + D \cos \sigma\tau, \qquad (6.3.19)$$

where σ is one of $\sigma_1, \ldots, \sigma_6$, and we look for a particular solution of the form

$$S = K_1 \sin \sigma\tau + L_1 \cos \sigma\tau + K_2 \tau \sin \sigma\tau + L_2 \tau \cos \sigma\tau, \qquad (6.3.20)$$

where K_1, L_1, K_2, and L_2 are constant symmetric tensors. When (6.3.19) and (6.3.20) are combined, we find that these tensors must satisfy the restrictions

$$\begin{aligned}
(\mathbb{L} - \sigma^2 \mathbb{M})K_1 &= \mathbb{M}(C + 2\sigma L_2), \\
(\mathbb{L} - \sigma^2 \mathbb{M})L_1 &= \mathbb{M}(D - 2\sigma K_2), \\
(\mathbb{L} - \sigma^2 \mathbb{M})K_2 &= 0, \\
(\mathbb{L} - \sigma^2 \mathbb{M})L_2 &= 0.
\end{aligned} \qquad (6.3.21)$$

Let us reorder the eigenvalues so that $\sigma^2 = \sigma_1^2 = \cdots = \sigma_r^2$ and $\{\sigma_{r+1}^2, \ldots,$ $\sigma_6^2\}$ does not contain σ^2. Then we see that

$$K_2 = \sum_{i=1}^{r} k_i A_i, \qquad L_2 = \sum_{i=1}^{r} l_i A_i \qquad (6.3.22)$$

for certain constants k_i and l_i, $i = 1, \ldots, r$. Next, we note that $(6.3.21)_1$ and $(6.3.21)_2$ are solvable if and only if the right-hand sides of these equations are orthogonal to $A_1, \ldots, A_r$ in the sense of (6.1.11):

$$A_i \cdot E_R^{1/2}(C + 2\sigma L_2)E_R^{1/2} = 0,$$
$$A_i \cdot E_R^{1/2}(D - 2\sigma K_2)E_R^{1/2} = 0, \qquad i = 1, \ldots, r. \qquad (6.3.23)$$

By (6.3.22) and the orthogonality relation (6.1.11), these imply that

$$k_i = \frac{1}{2\sigma} A_i \cdot E_R^{1/2} D E_R^{1/2},$$
$$\qquad\qquad\qquad\qquad\qquad\qquad i = 1, \ldots, r. \qquad (6.3.24)$$
$$l_i = -\frac{1}{2\sigma} A_i \cdot E_R^{1/2} C E_R^{1/2},$$

With these coefficients $(6.3.21)_1$ and $(6.3.21)_2$ are solvable, but the solution is unique only to within a linear combination of $A_1, \ldots, A_r$. Without loss of generality, we may look for a solution of the type

$$K_1 = \sum_{i=r+1}^{6} k_i A_i, \qquad L_1 = \sum_{i=r+1}^{6} l_i A_i. \qquad (6.3.25)$$

Placing these in $(6.3.21)_{1,2}$ and once again using (6.1.11), we deduce that

$$k_i = \frac{1}{\sigma_i^2 - \sigma^2} A_i \cdot E_R^{1/2} C E_R^{1/2},$$
$$\qquad\qquad\qquad\qquad\qquad\qquad i = r+1, \ldots, 6. \qquad (6.3.26)$$
$$l_i = \frac{1}{\sigma_i^2 - \sigma^2} A_i \cdot E_R^{1/2} D E_R^{1/2},$$

Collecting these results together, we find that a particular solution of (6.3.19) is given by

$$S = \sum_{r+1 \le i \le 6} \frac{1}{\sigma_i^2 - \sigma^2} ([A_i \cdot E_R^{1/2} C E_R^{1/2}] A_i \sin \sigma\tau + [A_i \cdot E_R^{1/2} D E_R^{1/2}] A_i \cos \sigma\tau)$$

$$+ \sum_{1 \le i \le r} \frac{1}{2\sigma} ([A_i \cdot E_R^{1/2} D E_R^{1/2}] A_i \tau \sin \sigma\tau - [A_i \cdot E_R^{1/2} C E_R^{1/2}] A_i \tau \cos \sigma\tau).$$
$$(6.3.27)$$

If σ is the eigenvalue σ_j, a more convenient form of this solution is

$$S = \sum_{i \notin I_j} \frac{1}{\sigma_i^2 - \sigma_j^2} ([A_i \cdot E_R^{1/2} C E_R^{1/2}] \sin \sigma_j\tau + [A_i \cdot E_R^{1/2} D E_R^{1/2}] \cos \sigma_j\tau) A_i$$

$$+ \sum_{i \in I_j} \frac{1}{2\sigma_j} ([A_i \cdot E_R^{1/2} D E_R^{1/2}] \tau \sin \sigma_j\tau - [A_i \cdot E_R^{1/2} C E_R^{1/2}] \tau \cos \sigma_j\tau) A_i.$$
$$(6.3.28)$$

In this form we can easily extend to the case of a nonhomogeneous equation involving several terms of the same general type as the right-hand side of (6.3.19). The result, when superimposed on (6.3.10) and the constant solution of (6.3.11), gives us the general solution (6.1.13). This completes the proof of the Spectral Theorem.

(iv) Special Results for Axially Symmetric Bodies

Pseudo-rigid bodies with an axis of symmetry play an important role in many applications. In particular, we consider in Section 6(v) the precession of a symmetric body that is almost rigid. In preparation for that analysis, we show here that for such bodies the eigenvalue problem (6.1.10) can be solved and the differential equations (6.1.20) displayed explicitly.

Recall that for a body with an axis of symmetry denoted by the unit vector D_1, E_R has the form (4.1.22) with respect to the tensor basis induced by an orthonormal basis D_i, $i = 1, 2, 3$. Here and in Section 6(v) we work exclusively in matrix notation, it being understood that the underlying basis is always that associated with the vectors D_i. Thus, we write

$$[E_R] = \text{diag}(E, E\delta, E\delta) \quad \text{or} \quad [E_R^{-1}] = \text{diag}(\xi, \xi\Delta, \xi\Delta), \qquad (6.4.1)$$

where $E > 0$, $\xi > 0$, $\delta > 0$, $\Delta > 0$, $\delta \neq 1$, $\Delta \neq 1$. If $[A] = (a_{ij})$, then (6.1.10) has the component form

$$\lambda\xi(a_{11} + a_{22} + a_{33}) + 2\mu\xi a_{11} = \sigma^2 a_{11},$$

$$\lambda\xi\Delta(a_{11} + a_{22} + a_{33}) + 2\mu\xi\Delta a_{22} = \sigma^2 a_{22},$$

$$\lambda\xi\Delta(a_{11} + a_{22} + a_{33}) + 2\mu\xi\Delta a_{33} = \sigma^2 a_{33},$$

$$\mu\xi(1 + \Delta)a_{12} = \sigma^2 a_{12}, \qquad (6.4.2)$$

$$\mu\xi(1 + \Delta)a_{13} = \sigma^2 a_{13},$$

$$2\mu\xi\Delta a_{23} = \sigma^2 a_{23},$$

where $\eta = \sigma^2$. Three eigenvalues and eigenvectors follow directly from the last three equations; namely,

$$\sigma_4^2 = 2\mu\xi\Delta, \qquad \sigma_5^2 = \mu\xi(1 + \Delta), \qquad \sigma_6^2 = \mu\xi(1 + \Delta), \qquad (6.4.3)$$

and

$$[A_4] = n_4 \begin{bmatrix} 0 & 0 & 0 \\ 0 & 0 & 1 \\ 0 & 1 & 0 \end{bmatrix}, \qquad [A_5] = n_5 \begin{bmatrix} 0 & 1 & 0 \\ 1 & 0 & 0 \\ 0 & 0 & 0 \end{bmatrix},$$

$$[A_6] = n_6 \begin{bmatrix} 0 & 0 & 1 \\ 0 & 0 & 0 \\ 1 & 0 & 0 \end{bmatrix}. \qquad (6.4.4)$$

The first three equations in (6.4.2) can easily be analyzed and give us three more solutions:

$$\sigma_1^2 = 2\mu\xi\Delta,$$

$$\sigma_2^2 = \frac{\xi}{2}[(2\Delta + 1)\lambda + 2\mu(\Delta + 1) + (((2\Delta + 1)\lambda$$

$$+ 2\mu(\Delta + 1))^2 - 8\mu\Delta(3\lambda + 2\mu))^{1/2}], \qquad (6.4.5)$$

$$\sigma_3^2 = \frac{\xi}{2}[(2\Delta + 1)\lambda + 2\mu(\Delta + 1) - (((2\Delta + 1)\lambda$$

$$+ 2\mu(\Delta + 1))^2 - 8\mu\Delta(3\lambda + 2\mu))^{1/2}],$$

and

$$[A_1] = n_1 \operatorname{diag}(0, 1, -1),$$

$$[A_2] = n_2 \operatorname{diag}(\sigma_2^2 - 2\Delta\xi(\lambda + \mu), \lambda\Delta\xi, \lambda\Delta\xi), \qquad (6.4.6)$$

$$[A_3] = n_3 \operatorname{diag}(\sigma_3^2 - 2\Delta\xi(\lambda + \mu), \lambda\Delta\xi, \lambda\Delta\xi).$$

The normalizing coefficients $n_1, \ldots, n_6$ are determined by (6.1.11), which now reduces to $A_i \cdot E_R^{1/2} A_i E_R^{1/2} = 1$ for $i = 1, \ldots, 6$. The computation of these coefficients is straightforward, the result being

$$n_1 = \left(\frac{\xi\Delta}{2}\right)^{1/2}, \qquad n_4 = \left(\frac{\xi\Delta}{2}\right)^{1/2}, \qquad n_5 = \left(\frac{\xi\sqrt{\Delta}}{2}\right)^{1/2}, \qquad n_6 = \left(\frac{\xi\sqrt{\Delta}}{2}\right)^{1/2},$$

$$n_2 = \frac{\sqrt{\xi}}{\sqrt{(\sigma_2^2 - 2\Delta\xi(\mu + \lambda))^2 + 2\lambda^2\Delta\xi^2}}, \qquad (6.4.7)$$

$$n_3 = \frac{\sqrt{\xi}}{\sqrt{(\sigma_3^2 - 2\Delta\xi(\mu + \lambda))^2 + 2\lambda^2\Delta\xi^2}}.$$

In order to find the functions p_{ji} appearing in the initial-value problems (6.1.20), we note the following intermediate results whose derivation, while sometimes tedious, is elementary:

$$E_R^{1/2} A_1 E_R^{1/2} = \frac{1}{\xi\Delta} A_1, \qquad E_R^{1/2} A_4 E_R^{1/2} = \frac{1}{\xi\Delta} A_4,$$

$$E_R^{1/2} A_5 E_R^{1/2} = \frac{1}{\xi\sqrt{\Delta}} A_5, \qquad E_R^{1/2} A_6 E_R^{1/2} = \frac{1}{\xi\sqrt{\Delta}} A_6,$$

$$(6.4.8)$$

$$[E_R^{1/2} A_2 E_R^{1/2}] = \frac{n_2}{\xi} \operatorname{diag}(\sigma_2^2 - 2\Delta\xi(\mu + \lambda), \lambda\xi, \lambda\xi),$$

$$[E_R^{1/2} A_3 E_R^{1/2}] = \frac{n_3}{\xi} \operatorname{diag}(\sigma_3^2 - 2\Delta\xi(\mu + \lambda), \lambda\xi, \lambda\xi).$$

Moreover, the quantities

$$Q_i = \text{sym}\left(\bar{\Omega}^{(0)} A_i + \frac{2\mu}{\sigma_i^2} \bar{\Omega}^{(0)} \, \text{sk}(A_i E_{\mathbf{R}}^{-1}) \right), \tag{6.4.9}$$

which also appear in the definition (6.1.19) of p_{ji}, can be computed explicitly. If $\bar{\Omega}^{(0)}$ is the skew tensor

$$[\bar{\Omega}^{(0)}] = \begin{bmatrix} 0 & -\Omega_3 & \Omega_2 \\ \Omega_3 & 0 & -\Omega_1 \\ -\Omega_2 & \Omega_1 & 0 \end{bmatrix}, \tag{6.4.10}$$

straightforward calculations give us

$$[Q_1] = \frac{n_1}{2}\begin{bmatrix} 0 & -\Omega_3 & -\Omega_2 \\ -\Omega_3 & 0 & 2\Omega_1 \\ -\Omega_2 & 2\Omega_1 & 0 \end{bmatrix}, \qquad [Q_4] = \frac{n_4}{2}\begin{bmatrix} 0 & \Omega_2 & -\Omega_3 \\ \Omega_2 & -2\Omega_1 & 0 \\ -\Omega_3 & 0 & 2\Omega_1 \end{bmatrix},$$

$$[Q_2] = \frac{n_2}{2}(\sigma_2^2 - \Delta\xi(3\lambda + 2\mu))\begin{bmatrix} 0 & \Omega_3 & -\Omega_2 \\ \Omega_3 & 0 & 0 \\ -\Omega_2 & 0 & 0 \end{bmatrix},$$

$$[Q_3] = \frac{n_3}{2}(\sigma_3^2 - \Delta\xi(3\lambda + 2\mu))\begin{bmatrix} 0 & \Omega_3 & -\Omega_2 \\ \Omega_3 & 0 & 0 \\ -\Omega_2 & 0 & 0 \end{bmatrix}, \tag{6.4.11}$$

$$[Q_5] = \frac{n_5}{1 + \Delta}\begin{bmatrix} -2\Omega_3 & 0 & \Omega_1 \\ 0 & 2\Delta\Omega_3 & -\Delta\Omega_2 \\ \Omega_1 & -\Delta\Omega_2 & 0 \end{bmatrix},$$

$$[Q_6] = \frac{n_6}{1 + \Delta}\begin{bmatrix} 2\Omega_2 & -\Omega_1 & 0 \\ -\Omega_1 & 0 & \Delta\Omega_3 \\ 0 & \Delta\Omega_3 & -2\Delta\Omega_2 \end{bmatrix}.$$

Using (6.4.8) and (6.4.11), we find that among the coefficients p_{ji} given by (6.1.19), the following particular values are those required in Section 6(v):

$$p_{ii} = 0, \qquad\qquad i = 1, \ldots, 6,$$

$$p_{14} = -\Omega_1, \qquad\qquad p_{41} = \Omega_1, \tag{6.4.12}$$

$$p_{56} = -\Omega_1/(1 + \Delta), \qquad p_{65} = \Omega_1/(1 + \Delta).$$

(v) Gyroscopic Motions

Gyroscopic motions conventionally pertain to the precession of a rigid body possessing an axis of symmetry. Within the framework of the transition to rigidity, we examine in this section analogous motions for a pseudo-rigid body that is almost rigid. As noted in the introductory remarks, this type of analysis addresses in at least a conceptually simple way the question of the flexibility of "satellites" and other bodies. Our aim is to determine how the effects of flexibility alter the motion of a body that is largely exhibiting a rigid-body precession.

Let the axis of symmetry of the body be given by a unit vector D_1 associated with an orthonormal basis D_1, D_2, D_3 fixed in the reference placement. Relative to this basis, the Euler tensor of the body has the form (6.4.1). For convenience we write s for D_1. The classical solution in rigid-body mechanics for a body undergoing precession is given by a rotation tensor having the form (cf. Section 4(i))

$$R(t) = R_p(t)R_s(t) = \exp(\dot{p}tP)\exp(\dot{s}tS), \qquad (6.5.1)$$

in which $\dot{p}$ and $\dot{s}$ are constants related by

$$\dot{s} = \frac{\Delta - 1}{2}\dot{p}\cos\alpha, \qquad (6.5.2)$$

S is a skew tensor with axis s, and P is the skew tensor with axis p given by $(4.1.23)_1$. In matrix form,

$$[P] = \begin{bmatrix} 0 & \sin\alpha & 0 \\ -\sin\alpha & 0 & -\cos\alpha \\ 0 & \cos\alpha & 0 \end{bmatrix}, \qquad [S] = \begin{bmatrix} 0 & 0 & 0 \\ 0 & 0 & -1 \\ 0 & 1 & 0 \end{bmatrix}. \quad (6.5.3)$$

Physically, the body is rotating about the axis s with constant angular velocity $\dot{s}$ while that axis is rotating about the axis p with constant angular velocity $\dot{p}$. The angle between these two axes is α.

We wish to use (6.5.1) as the leading term in our analysis of perturbations for an almost-rigid body. This classical solution is appropriate only when there are no external torques present. For the current application, we assume more generally that all external effects vanish and, in particular, that $\overline{M}(t) = 0$.

Let $\overline{\Omega}^{(0)} = R^T\dot{R}$. A simple calculation then gives us

$$\overline{\Omega}^{(0)} = R_s^T(\dot{p}P + \dot{s}S)R_s,$$
$$\dot{\overline{\Omega}}^{(0)} + \overline{\Omega}^{(0)2} = R_s^T(\dot{p}^2P^2 + 2\dot{p}\dot{s}PS + \dot{s}^2S^2)R_s. \qquad (6.5.4)$$

Using (6.5.3), we can easily compute the components of $\overline{\Omega}^{(0)}$ explicitly. In terms of the general representation (6.4.10), we obtain

$$\Omega_1 = \dot{p}\cos\alpha + \dot{s} = \frac{1+\Delta}{2}\dot{p}\cos\alpha \equiv \Omega,$$

$$\Omega_2 = -\dot{p}\sin\alpha\sin\dot{s}t, \qquad (6.5.5)$$

$$\Omega_3 = -\dot{p}\sin\alpha\cos\dot{s}t,$$

where in the first line we have used (6.5.2) to eliminate $\dot{s}$.

The representations (6.4.1) and (6.5.3) show that $(\overset{\star}{\bar{\Omega}}{}^{(0)} - \bar{\Omega}^{(0)2})E_R$ is a symmetric tensor provided that $\dot{s}$ and $\dot{p}$ are related through (6.5.2). Therefore, $(6.1.16)_1$ of the Transition Theorem is satisfied. Moreover, this symmetric tensor determines the initial approximation $\bar{S}^{(0)}$ of the strains through (6.1.18); that equation is given explicitly in the present case by

$$[\lambda(\operatorname{tr} \bar{S}^{(0)})I + 2\mu\bar{S}^{(0)}] =$$

$$\frac{\dot{p}^2}{\xi\Delta}[R_s^T] \begin{bmatrix} \Delta \sin^2 \alpha & 0 & \Delta \sin \alpha \cos \alpha \\ 0 & \sin^2 \alpha + \dfrac{(\Delta + 1)^2}{4}\cos^2 \alpha & 0 \\ \Delta \sin \alpha \cos \alpha & 0 & \dfrac{(\Delta + 1)^2}{4}\cos^2 \alpha \end{bmatrix} [R_s].$$

$$\tag{6.5.6}$$

Solving for $\bar{S}^{(0)}$, we obtain

$$[\bar{S}^{(0)}] = \frac{\dot{p}^2}{\xi\Delta\mu(3\lambda + 2\mu)}[R_s^T] \begin{bmatrix} G_{11} & 0 & G_{13} \\ 0 & G_{22} & 0 \\ G_{13} & 0 & G_{33} \end{bmatrix} [R_s], \tag{6.5.7}$$

where

$$G_{11} = 4(2\Delta(\lambda + \mu) - \lambda) \sin^2 \alpha - 2\lambda(\Delta + 1)^2 \cos^2 \alpha,$$
$$G_{22} = 4(2(\lambda + \mu) - \lambda\Delta) \sin^2 \alpha + (\lambda + 2\mu)(\Delta + 1)^2 \cos^2 \alpha,$$
$$G_{33} = -4\lambda(\Delta + 1) \sin^2 \alpha + (\lambda + 2\mu)(\Delta + 1)^2 \cos^2 \alpha, \tag{6.5.8}$$
$$G_{13} = 4(3\lambda + 2\mu)\Delta \sin \alpha \cos \alpha.$$

The right-hand side of (6.5.6) is essentially the external force-moment presented in (4.1.40), namely, the force-moment that must be imposed for a pseudo-rigid body to execute *exactly* a rigid motion. In the present case, there is no external force-moment, and so the contribution it made previously must now be balanced, as expressed in (6.5.6), by force-moments resulting from internal strains. It is interesting to note from (6.5.7) that these strains represent expansions and contractions along the basic triad D_1, D_2, and D_3 and a shear in the plane of the two axes of rotation p and s.

Consider next the terms $\bar{\Omega}^{(1)}$ and $\bar{S}^{(0)}$ in the general expansions (6.1.15). By (6.1.21) these are determined by the functions $\alpha_i(t)$ and $\beta_i(t)$, and thus we must solve the initial-value problems (6.1.20). For an axially symmetric body, the coefficients p_{ij} are given by (6.4.12). Using these expressions and the facts that $\sigma_1^2 = \sigma_4^2$ and $\sigma_5^2 = \sigma_6^2$, we obtain for α_i the system

$$\dot{\alpha}_1 - \Omega\alpha_4 = 0,$$

$$\dot{\alpha}_4 + \Omega\alpha_1 = 0,$$

$$\dot{\alpha}_2 = 0,$$

$$\dot{\alpha}_3 = 0,$$

$$\dot{\alpha}_5 - \frac{\Omega}{1+\Delta}\alpha_6 = 0,$$

$$\dot{\alpha}_6 + \frac{\Omega}{1+\Delta}\alpha_5 = 0.$$

(6.5.9)

By $(6.5.5)_1$, Ω is a constant. Therefore this is a linear system with constant coefficients and it can be solved by standard techniques. We obtain

$$\alpha_1(t) = \alpha_{11} \cos \Omega t + \alpha_{12} \sin \Omega t,$$

$$\alpha_4(t) = \alpha_{12} \cos \Omega t - \alpha_{11} \sin \Omega t,$$

$$\alpha_2(t) = \alpha_{21}, \qquad \alpha_3(t) = \alpha_{31},$$

$$\alpha_5(t) = \alpha_{51} \cos\left(\frac{\Omega}{1+\Delta}t\right) + \alpha_{52} \sin\left(\frac{\Omega}{1+\Delta}t\right),$$

$$\alpha_6(t) = \alpha_{52} \cos\left(\frac{\Omega}{1+\Delta}t\right) - \alpha_{51} \sin\left(\frac{\Omega}{1+\Delta}t\right),$$

(6.5.10)

where α_{11}, α_{12}, ..., α_{52} are constants of integration. The functions $\beta_i(t)$ have the same form, though with different coefficients. These results and (6.1.21) show that the contributions to the spin and the strain owing to $\hat{\Omega}^{(1)}$ and $S^{(0)}$ have the form of high-frequency oscillations whose amplitude varies periodically in time on the slow time scale. This effect is analogous to the occurrence of "beats" in a forced harmonic oscillator. At the same time, we note that such beats do not appear in the component of spin along the axis of symmetry s. Indeed, using the explicit representations (6.4.4) and (6.4.6) for the tensors A_i, we find that

$$\text{sk}(A_i E_R^{-1}) = 0 \quad \text{for } i = 1, 2, 3, 4,$$

$$[\text{sk}(A_5 E_R^{-1})] = \frac{\xi(\Delta - 1)}{2}\begin{bmatrix} 0 & 1 & 0 \\ -1 & 0 & 0 \\ 0 & 0 & 0 \end{bmatrix},$$

(6.5.11)

$$[\text{sk}(A_6 E_R^{-1})] = \frac{\xi(\Delta - 1)}{2}\begin{bmatrix} 0 & 0 & 1 \\ 0 & 0 & 0 \\ -1 & 0 & 0 \end{bmatrix},$$

and so the entry in the second row and third column of $[\hat{\Omega}^{(1)}]$ vanishes.

Finally, we consider the first correction $\bar{\mathbf{\Omega}}^{(1)}$ to $\mathbf{\Omega}^{\varepsilon}$. Setting

$$[\bar{\mathbf{\Omega}}^{(1)}] = \begin{bmatrix} 0 & -\Lambda_3 & \Lambda_2 \\ \Lambda_3 & 0 & -\Lambda_1 \\ -\Lambda_2 & \Lambda_1 & 0 \end{bmatrix} \tag{6.5.12}$$

and recalling the general expression (6.4.10) for $[\bar{\mathbf{\Omega}}^{(0)}]$, we obtain the following explicit form of the system $(6.1.17)_1$:

$$\dot{\Lambda}_1 = 0,$$

$$\dot{\Lambda}_2 + \frac{1-\Delta}{1+\Delta}(\Omega_3\Lambda_1 + \Omega_1\Lambda_3) = 0, \tag{6.5.13}$$

$$\dot{\Lambda}_3 - \frac{1-\Delta}{1+\Delta}(\Omega_2\Lambda_1 + \Omega_1\Lambda_2) = 0.$$

The initial conditions for Λ are determined by the strain-rate tensor L through $(6.1.17)_2$. Using (6.4.3)–(6.4.8), (6.5.11), and (6.5.12), we can write these as

$$\Lambda_1(0) = 0,$$

$$\Lambda_2(0) = \frac{1-\Delta}{1+\Delta}\left(\frac{2}{\xi\sqrt{\Delta}}\right)^{1/2} L_{13}, \tag{6.5.14}$$

$$\Lambda_3(0) = -\frac{1-\Delta}{1+\Delta}\left(\frac{2}{\xi\sqrt{\Delta}}\right)^{1/2} L_{12}.$$

By $(6.5.13)_1$ and $(6.5.14)_1$, we see that

$$\Lambda_1(t) = 0. \tag{6.5.15}$$

Moreover, since Ω_1 is the constant Ω, the system for Λ_2 and Λ_3 is linear with constant coefficients. Its solution is

$$\Lambda_2(t) = A\cos\left(\frac{1-\Delta}{1+\Delta}\Omega t\right) + B\sin\left(\frac{1-\Delta}{1+\Delta}\Omega t\right),$$

$$\Lambda_3(t) = -B\cos\left(\frac{1-\Delta}{1+\Delta}\Omega t\right) + A\sin\left(\frac{1-\Delta}{1+\Delta}\Omega t\right), \tag{6.5.16}$$

where $A = \Lambda_2(0)$ and $B = -\Lambda_3(0)$, these being given by (6.5.14).

In order to interpret this correction to the angular velocity, we note first from (6.5.5) and (6.5.2) that

$$\frac{1-\Delta}{1+\Delta}\Omega = -\dot{s}. \tag{6.5.17}$$

Therefore, the components of $\omega = \text{ax}(\bar{\mathbf{\Omega}}^{(0)} + \varepsilon\bar{\mathbf{\Omega}}^{(1)})$ have the form

$\omega_1 = \Omega,$

$\omega_2 = \varepsilon A \cos \dot{s}t - (\varepsilon B + \dot{p} \sin \alpha) \sin \dot{s}t = -K \sin(\dot{s}t + \delta),$ \qquad (6.5.18)

$\omega_3 = -(\varepsilon B + \dot{p} \sin \alpha) \cos \dot{s}t - \varepsilon A \sin \dot{s}t = -K \cos(\dot{s}t + \delta),$

where we have set

$$K = (\varepsilon^2 A^2 + (\varepsilon B + \dot{p} \sin \alpha)^2)^{1/2},$$

$$\cos \delta = (\varepsilon B + \dot{p} \sin \alpha)/K, \qquad (6.5.19)$$

$$\sin \delta = -\varepsilon A/K.$$

The perturbations to $\boldsymbol{\Omega}^\varepsilon$ contributed by Λ_2 and Λ_3 have the effect of leaving the body in a state of precession, though with the new amplitude given by K and with the lag in time signified by the angle δ. These perturbations are the result of the nonzero components L_{12} and L_{13} of the strain-rate tensor L, rates that create shears in the body in planes that contain the axis of symmetry s.

It is important to note that the strain $S^{(0)}$ contributes to the rotation only through the amplitude-modulated term $\hat{\boldsymbol{\Omega}}^{(1)}$. By appropriate choices of the initial strain K and strain-rate L, it is possible to make this term vanish. It is also evident that the calculation of the perturbation $\boldsymbol{\Omega}^{(2)}$ will bring to bear the full effect of $S^{(0)}$ on the rotation. An assessment of the relative importance of this correction must await further study.

References

Listed here are works we have found useful in our development and presentation of the theory of pseudo-rigid bodies. It is not a comprehensive bibliography. The number of works bearing directly upon our subject is currently small. A wealth of others are related either by analogy or slight overlap; we have included only those that we feel make a substantial contribution.

[1893] P. DUHEM, Le potential thermodynamique et la pression hydrostatic, *Ann. École Norm.* **10**(3): 187–230.

[1909] E. COSSERAT and F. COSSERAT, *Theorie des Corps Deformables*, Hermann et Fils, Paris.

[1918] E. NOETHER, Invariante variationsprobleme, *Nachr. Ges. Wiss. Göttingen Math.-Phys. Kl.* 235–257.

[1927] A. E. H. LOVE, *The Mathematical Theory of Elasticity*, Historical Introduction, Cambridge University Press, Cambridge, England.

[1954] J. A. SCHOUTEN, *Ricci-Calculus*, 2nd ed., Springer, Berlin.

[1958] J. L. ERICKSEN and C. TRUESDELL, Exact theory of stress and strain in rods and shells, *Arch. Rational Mech. Anal.* **1**: 295–323.

[1959] J. L. SYNGE, *Principles of Mechanics*, 3rd ed., McGraw-Hill, New York.
       C. TRUESDELL, The rational mechanics of materials—past, present, future, *Appl. Mech. Rev.* **12**: 75–80.

[1960] F. JOHN, Plane strain problems for a perfectly elastic material of harmonic type, *Commun. Pure Appl. Math.* **13**: 239–296.
       K. R. SYMON, *Mechanics*, 2nd ed., Addison-Wesley, Reading, MA.
       J. L. SYNGE, Classical dynamics, in *Handbuch der Physik*, Vol. III/1 (S. Flügge, ed.), Springer, Berlin.
       C. TRUESDELL and R. A. TOUPIN, The classical field theories, in *Handbuch der Physik*, Vol. III/1 (S. Flügge, ed.), Springer, Berlin.

[1962] P. J. BLATZ and W. L. KO, Application of finite elasticity theory to the deformation of rubbery materials, *Trans. Soc. Rheol.* **6**: 223–251.

[1964] R. D. MINDLIN, Microstructure in linear elasticity, *Arch. Rational Mech. Anal.* **16**: 51–78.
       W. NOLL, Euclidean geometry and Minkowskian chronometry, *Am. Math. Monthly* **71**: 129–144.
       R. A. TOUPIN, Theories of elasticity with couple stress, *Arch. Rational Mech. Anal.* **17**: 85–112.

[1965] D. T. GREENWOOD, *Principles of Dynamics*, Prentice-Hall, Englewood Cliffs, NJ.

L. A. PARS, *A Treatise on Analytical Dynamics*, Wiley, New York.

C. TRUESDELL and W. NOLL, The non-linear field theories of mechanics, in *Handbuch der Physik*, Vol. III/3 (S. Flügge, ed.), Springer, Berlin.

[1966] H. COHEN and C. N. DESILVA, Nonlinear theory of elastic directed surfaces, *J. Math. Phys.* **7**: 960–966.

H. COHEN, A non-linear theory of elastic directed curves, *Int. J. Eng. Sci.* **4**: 511–524.

[1967] W. JAUNZEMIS, *Continuum Mechanics*, Macmillan, New York.

[1968] E. KRONER (ed.), *Mechanics of Generalized Continua, Proceedings of an IUTAM Symposium*, Fruedenstadt and Stuttgart, 1967, Springer, Berlin.

A. I. LURE, Theory of elasticity for a semi-linear material, *J. Appl. Math. Mech.* **32**: 1068–1085.

[1969] J. K. HALE, *Ordinary Differential Equations*, Wiley, New York.

[1970] P. N. KALONI and C. N. DESILVA, A theory of oriented fluids, *Phys. Fluids* **13**: 1708–1716.

[1971] E. J. SALETAN and A. H. CROMER, *Theoretical Mechanics*, Wiley, New York.

[1972] S. S. ANTMAN, The theory of rods, in *Handbuch der Physik*, Vol. VIa/2 (C. Truesdell, ed.), Springer, Berlin.

P. M. NAGHDI, The theory of shells, in *Handbuch der Physik*, Vol. VIa/2 (C. Truesdell, ed.), Springer, Berlin.

[1973] C.-C. WANG and C. TRUESDELL, *Introduction to Rational Elasticity*, Noordhoff International, Leyden.

[1974] J. J. SLAWIANOWSKI, Analytical mechanics of finite homogeneous strains, *Arch. Mech.* **26**: 569–587.

[1975] J. J. SLAWIANOWSKI, Newtonian dynamics of homogeneous strains, *Arch. Mech.* **27**: 93–102.

[1976] G. CAPRIZ and P. PODIO-GUIDUGLI, Discrete and continuous bodies with affine structure, *Ann. Mat. Pura Appl.* **111**: 195–217.

C. J. H. WILLIAMS, Dynamic modelling and formulation techniques for non-rigid spacecraft, in *Proceedings E. S. A. Symposium on Dynamics and Control of Non-rigid Spacecraft*, ESA SP 117, Frascati, Italy.

[1977] T. J. R. HUGHES, T. KATO, and J. E. MARSDEN, Well-posed quasi-linear second-order hyperbolic systems with applications to nonlinear elastodynamics and general relativity, *Arch. Rational Mech. Anal.* **63**: 273–294.

C. TRUESDELL, *A First Course in Rational Continuum Mechanics*, Vol. 1, Academic, New York.

[1978] M. D. GREENBERG, *Foundations of Applied Mathematics*, Prentice-Hall, Englewood Cliffs, NJ.

[1979] G. DEL PIERO, Some properties of the set of fourth-order tensors, with application to elasticity, *J. Elasticity* **9**: 245–261.

C.-C. WANG, *Mathematical Principles of Mechanics and Electrodynamics, Part A: Analytical and Continuum Mechanics*, Plenum, New York.

[1980] H. GOLDSTEIN, *Classical Mechanics*, 2nd ed., Addison-Wesley, Reading, MA.

K. LAMBECK, *The Earth's Variable Rotation: Geophysical Causes and Consequences*, Cambridge University Press, Cambridge, England.

C. TRUESDELL, Sketch for a history of constitutive relations, in *Rheology*, Vol. 1 (G. Astarita, G. Marruci, and L. Nicolais, eds.), Plenum, New York.

C. TRUESDELL and R. G. MUNCASTER, *Fundamentals of Maxwell's Kinetic Theory of a Simple Monatomic Gas*, Academic, New York.

[1981]   H. COHEN, Pseudo-rigid bodies, *Utilitas Math.* **20**: 221–247.

[1982]   O. BRULIN and R. K. T. HSIEH (eds.), *Mechanics of Micropolar Media,* *CISM Lectures,* Udine, 1979, World Scientific, Singapore.

[1983]   J. E. MARSDEN and T. J. R. HUGHES, *Mathematical Foundations of Elasticity,* Prentice-Hall, Englewood Cliffs, NJ.

[1984]   H. COHEN and R. G. MUNCASTER, The dynamics of pseudo-rigid bodies: General structure and exact solutions, *J. Elasticity* **14**: 127–154.

H. COHEN and C.-C. WANG, On the response and symmetry of elastic and hyperelastic membrane points, *Arch. Rational Mech. Anal.* **85**: 355–391.

a. R. G. MUNCASTER, Invariant manifolds in mechanics I: The general construction of coarse theories from fine theories, *Arch. Rational Mech. Anal.* **84**: 353–373.

b. R. G. MUNCASTER, Invariant manifolds in mechanics II: Zero-dimensional elastic bodies with directors, *Arch. Rational Mech. Anal.* **84**:375–392.

H. C. SIMPSON and S. J. SPECTOR, On barreling instability in nonlinear elasticity, *J. Elasticity* **14**: 103–125.

[1985]   J. PIERCE, Global models for Cosserat continua and some fibrations of Palais, Cerf, and Smale, in *Physical Mathematics and Nonlinear Differential Equations* (J. H. Lightbourne III and S. M. Rankin III, eds.), Marcel Dekker, New York.

[1986]   H. COHEN and R. G. MUNCASTER, Rolling of a pseudo-rigid lamina, in *SECTAM XIII Proceedings,* Columbia, SC: 77–81.

[1988]   H. COHEN and G. MACSITHIGH, Plane motions of elastic pseudo-rigid bodies, *J. Elasticity* (to appear).

T. HEALEY, Symmetry and equivariance in nonlinear elastostatics I: Differential field equations, *Arch. Rational Mech. Anal.* (to appear).

Index